FAO中文出版计划项目丛书

土壤污染：一个隐藏的现实

联合国粮食及农业组织 编著
陈保青 刘海涛 译

中国农业出版社
联合国粮食及农业组织
2021·北京

土壤污染：一个隐藏的现实

引用格式要求：

粮农组织和中国农业出版社。2021年。《土壤污染：一个隐藏的现实》。中国北京。

19-CPP2020

本出版物原版为英文，即 *Soil Pollution: a hidden reality*，由联合国粮食及农业组织于2018年出版。此中文翻译由中国农业科学院农业环境与可持续发展研究所安排并对翻译的准确性及质量负全部责任。如有出入，应以英文原版为准。

本信息产品中使用的名称和介绍的材料，并不意味着联合国粮食及农业组织（粮农组织）对任何国家、领地、城市、地区或其当局的法律或发展状况，或对其国界或边界的划分表示任何意见。提及具体的公司或厂商产品，无论是否含有专利，并不意味着这些公司或产品得到粮农组织的认可或推荐，优于未提及的其他类似公司或产品。

本信息产品中陈述的观点是作者的观点，不一定反映粮农组织的观点或政策。

ISBN 978-92-5-134698-3（粮农组织）
ISBN 978-7-109-28113-4（中国农业出版社）

© 粮农组织，2018年（英文版）
© 粮农组织，2021年（中文版）

保留部分权利。本作品根据署名-非商业性使用-相同方式共享3.0政府间组织许可（CC BY-NC- SA 3.0 IGO; https://creativecommons.org/licenses/by-nc-sa/3.0/igo）公开。

根据该许可条款，本作品可被复制、再次传播和改编，以用于非商业目的，但必须恰当引用。使用本作品时不应暗示粮农组织认可任何具体的组织、产品或服务。不允许使用粮农组织标识。如对本作品进行改编，则必须获得相同或等效的知识共享许可。如翻译本作品，必须包含所要求的引用和下述免责声明："该译文并非由联合国粮食及农业组织（粮农组织）生成。粮农组织不对本翻译的内容或准确性负责。原英文版本应为权威版本。"

除非另有规定，本许可下产生的争议，如通过调解无法友好解决，则按本许可第8条之规定，通过仲裁解决。适用的调解规则为世界知识产权组织调解规则（http://www.wipo.int/amc/en/mediation/rules），任何仲裁将遵循联合国国际贸易法委员会（贸法委）的仲裁规则进行仲裁。

第三方材料。 欲再利用本作品中属于第三方的材料（如表格、图形或图片）的用户，需自行判断再利用是否需要许可，并自行向版权持有者申请许可。对任何第三方所有的材料侵权而导致的索赔风险完全由用户承担。

销售、权利和授权。 粮农组织信息产品可在粮农组织网站（www.fao.org/publications）获得，也可通过publications-sales@fao.org购买。商业性使用的申请应递交至www.fao.org/contact-us/licence-request。关于权利和授权的征询应递交至copyright@fao.org。

FAO中文出版计划项目丛书

指导委员会

主　任　隋鹏飞
副主任　谢建民　倪洪兴　韦正林　彭廷军　童玉娥　蔺惠芳
委　员　徐　明　徐玉波　朱宝颖　傅永东

FAO中文出版计划项目丛书

译审委员会

主　任　童玉娥　蔺惠芳

副主任　罗　鸣　苑　荣　刘爱芳　徐　明

编　委　闫保荣　刘雁南　宋雨星　安　全　张夕珺
　　　　　李巧巧　宋　莉　赵　文　刘海涛　黄　波
　　　　　张龙豹　朱亚勤　李　熙　郑　君

本书译审名单

翻　译　陈保青　刘海涛

审　校　刘海涛　陈保青　张龙豹

概　要

"土壤污染"是指土壤中的化学品或物质出现的位置不当或浓度高于正常水平，对非靶向生物会产生不利影响。土壤污染往往无法直接评估或直观感知，从而成为一种隐患。

《世界土壤资源状况报告》指出，土壤污染是影响全球土壤及其生态系统服务的主要土壤威胁之一。

每个地区对土壤污染的关注程度都在增加。最近，联合国环境大会(UNEA-3)通过了一项决议，呼吁加快行动和合作，以解决和管理土壤污染。170多个国家达成的这一共识清楚地表明，土壤污染已具有全球相关性，也表明这些国家具有为解决污染问题而制定具体解决方案的意愿。

土壤污染的主要人为来源是工业活动中所使用的或产生的化学品和副产物，如家用、牲畜和市政废物(包括废水)，农用化学品和石油衍生产品。这些化学物质被无意（例如来自石油泄漏或垃圾填埋场的沥滤）或有意地（例如使用化肥和农药、用未经处理的废水灌溉或在土地上施用污水污泥）释放到环境中。土壤污染也可以来源于由冶炼、运输、喷洒农药造成的大气沉降，物质的不完全燃烧，以及由武器试验和核事故造成的放射性核素的大气沉降。药物、内分泌干扰物、激素和毒素，以及包括细菌和病毒生物污染在内的土壤微生物污染物等新型污染物也引起了人们新的关注。

科学证据表明，土壤污染会严重降低土壤提供的主要生态系统服务。土壤污染降低了食品安全，一方面污染物的毒性程度降低了作物产量；另一方面在污染土壤中生产的作物已不能供动物和人类安全食用。许多污染物(包括氮、磷等主要营养物质)从土壤中转移到地表水和地下水中，导致水体富营养化，对环境造成巨大危害，且受污染的水被饮用后会直接影响人类健康。污染物还直接损害土壤微生物和栖居在土壤中的生物，从而影响土壤生物多样性以及该生物体所产生的生态功能。

科学研究结果表明，土壤污染直接影响人体健康。砷、铅和镉等元素，多氯联苯和多环芳香烃等有机化学品，以及抗生素等药品的污染，都对人类健康构成了威胁。1986年切尔诺贝利核事故造成的大面积土壤放射性核素污染对许多人来说都是一个永恒的记忆。

污染土壤的修复是必不可少的，发展新的、以科学为基础的修复方法的相关研究也在继续进行。全世界的风险评估方法都是类似的，包括采取一系列步骤以确定和评价是自然物质还是人为物质造成的土壤污染,以及这种污染对环境和人类健康构成风险的程度。日益昂贵的物理修复方法(如垃圾填埋场中的化学灭活或隔离)正在被基于科学的生物方法(如微生物降解或植物修复方法)所代替。

联合国粮食及农业组织(FAO)修订的《世界土壤宪章》建议国家政府简化土壤污染管理条例,制定污染物积累限定水平,以提供健康的环境和安全的食品,保障人类健康和福祉。各国政府还被敦促协助修复超过限定浓度的污染土壤,在全球执行可持续的土壤管理措施,限制农业来源的污染。

本书旨在总结土壤污染的现状,并确定影响人类健康和环境的主要污染物及其来源,并对那些存在于农业系统并通过食物链到达人体的污染物进行特殊关注。最后对评估和修复污染土壤的最佳可行技术进行了一些个案总结。

本书在全球土壤污染研讨会(GSOP18)的基础上进行撰写,明确了世界范围内土壤污染知识的主要差距,为今后的进一步研究提供了依据。

术语表

污染物：由于人类活动而存在于土壤中的物质〔国际标准化组织(ISO)，2013〕。

淋溶：水溶性物质的溶解和运动（ISO，2013）。

母质：土壤形成过程中的原始物质（矿物质和有机物）。

持久性有机污染物（POPs）：农业化学品和工业产品中合成的碳基化合物，一般生物降解性极差，大部分可以在生物组织中积累。一些农药属于持久性有机污染物，如多氯二苯并二噁英(PCDDs)、多氯二苯并呋喃(PCDFs)、多氯联苯(PCBs)和多环芳香烃(PAHs)。

土壤：由于风化、物理、化学和生物过程而发生转化的地壳上层。它由矿物颗粒、有机物、水、空气和土壤发生层中的活体组织组成（ISO，2013）。

土壤生态系统功能：土壤对于人类和环境重要性的描述。如：①控制着生态系统内物质和能量循环；②是植物、动物和人类的生活基础；③是建筑物、道路的稳定基础；④是农业、林业发展的基础；⑤是遗传库载体；⑥是自然史记载；⑦是考古学和古生态记载（ISO，2013）。

土壤健康：土壤在生态系统中和一定的土地使用范围内，作为一个重要的生命系统持续发挥作用以维持生物生产力，提高空气和水环境的质量，维持植物、动物和人类的健康的能力（Doran, Stamatiadis 和 Haberern, 2002）。

土壤生态系统服务：土壤中发生的自然过程和存在的组分直接或间接地提供满足人类需要的商品和服务(Groot, 1992)。

食品安全：它被定义为食物的可获取性、流通性、可利用性和供应的稳定性。

土壤污染物：土壤中出现的一种化学品或物质，其浓度高于自然条件下本该有的浓度，但是在此浓度下不一定会造成伤害。

土壤污染：指化学品或物质出现位置不当或浓度高于正常水平，对非靶向生物会产生不利影响。

目 录

概要 ··· v
术语表 ··· vii

1 什么是土壤污染？ ·· 1
1.1 引言 ·· 1
1.2 点源和土壤污染扩散 ··· 3
1.2.1 点源污染 ·· 4
1.2.2 扩散污染 ·· 4
1.3 土壤污染来源 ·· 6
1.3.1 自然、地球成因来源 ·································· 6
1.3.2 人为源 ·· 8
1.4 土壤主要污染物 ·· 18
1.4.1 重金属和类金属 ·· 18
1.4.2 氮和磷 ·· 19
1.4.3 农药 ·· 20
1.4.4 多环芳烃 ·· 23
1.4.5 持久性有机污染物 ···································· 25
1.4.6 放射性核素 ·· 28
1.4.7 新型污染物 ·· 29
1.4.8 病原微生物 ·· 33
1.4.9 抗生素抗性细菌和基因 ···························· 34

1.5 污染物与土壤成分的相互作用 ………………………………… 35
    1.5.1 污染物的吸附 ……………………………………………… 36
    1.5.2 污染物的生物有效性、流动性和降解 ………………… 37

2 土壤污染对食物链和生态系统服务的影响 …………………… 41
2.1 土壤污染、植物吸收和食物链污染 ……………………………… 42
2.2 农业土壤污染对生态系统服务的影响 …………………………… 45
    2.2.1 化学肥料 …………………………………………………… 45
    2.2.2 酸化和作物损失 ………………………………………… 45
    2.2.3 农药 ………………………………………………………… 46
    2.2.4 粪肥 ………………………………………………………… 46
    2.2.5 农业生产中的城市废弃物 ……………………………… 47
2.3 与土壤污染有关的人类健康风险 ………………………………… 48
    2.3.1 人类接触土壤污染物的途径及其对人类健康的影响 … 48
    2.3.2 土壤是抗生素耐药性细菌和基因的储存库 …………… 53

3 污染土壤的管理和修复 …………………………………………… 57
3.1 风险评估方法 ………………………………………………………… 57
3.2 污染场地修复的主要技术 ………………………………………… 63
3.3 降低食物链污染和生态系统服务的农业生产方式改变 …… 67
    3.3.1 化肥 ………………………………………………………… 67
    3.3.2 农药 ………………………………………………………… 69
    3.3.3 金属 ………………………………………………………… 70
    3.3.4 类金属 ……………………………………………………… 71
    3.3.5 放射性核素 ……………………………………………… 71

4 土壤污染与修复案例研究 ………………………………………… 73
4.1 联合国外地特派团通过加强自然衰减来进行POL污染地的
     修复：关于联合国在科特迪瓦行动的个案研究 ……………… 73

4.2 西伯利亚西部针叶林带利用当代修复方法进行石油污染土地的
修复 ·· 74
4.3 辅助植物固定：西班牙东南部尾矿的一种
有效修复技术 ··· 76

参考文献 ··· 78

1 什么是土壤污染？

1.1 引言

"土壤污染"是指化学品或其他物质在土壤中出现位置不当，或浓度高于正常水平，对非靶向生物会产生不利影响[联合国粮食及农业组织（FAO）和全球土壤合作伙伴关系政府间土壤技术专家组（ITPS），2015]。虽然大多数污染物是人为产生的，但有些污染物作为矿物成分可以在土壤中自然地产生，且其在高浓度下可能是有毒的。土壤污染往往无法被直接评估或直观感知，因此成为一种隐患。

由于农业化学和工业的发展，污染物的种类在不断变化。这种多样性以及土壤中的有机物通过生物活动产生不同的代谢物，使得通过土壤调查来识别污染物既困难又昂贵。与此同时，土壤污染的影响也取决于土壤特性，因为这些特性控制着污染物的迁移性、生物可利用程度和污染持续时间（FAO和ITPS，2015）。

工业化、战争、矿业和农业集约化给世界各地带来了土壤污染的后遗症[Bundschuh等，2012；环境事务部（DEA），2010；欧洲环境署（EEA），2014；Luo等，2009；英国SSR公司（SSR），2010]。自城市化扩张以来，土壤被用作倾倒固体和液体废物的场所。人们认为，污染物一旦掩埋和消失不见就将不会对人类健康或环境构成任何威胁，并且它们将以某种方式消失（Swartjes，2011）。土壤污染的主要来源是人为来源，人为污染导致污染物在土壤中累积，达到可以引起人们关注的程度（Cachada，Rocha-Santos和Duarte，2018）。

土壤污染是一个令人担忧的问题。土壤污染被认为是欧洲和欧亚大陆土壤功能的第三大威胁，是北非第四、亚洲第五、西北太平洋第七、北美第八、撒哈拉以南非洲和拉丁美洲第九大威胁（FAO和ITPS，2015）。某些污染物的存在还可能造成土壤养分不均衡和土壤酸化，正如《世界土壤资源状况报告》所指出的那样，这是世界许多地区的两大主要问题（FAO和ITPS，2015）。

在20世纪90年代，国际土壤参考和信息中心（International Soil Reference and Information Centre，ISRIC）和联合国环境规划署（United Nations Environment Programme，UNEP）对全球土壤污染进行了特殊评估，据估计有2 200万公顷土地已经受到了土壤污染的影响（Oldeman，1991）。然而，最新数据表明，他们可能低估了问题的严重性。因为国家尺度的土壤污染程度评估主要是在发达国家中进行的。而根据中国环境保护部的数据,中国有16%的土壤和19%的农业土壤被列为污染土壤[中国环境与发展国际合作委员会(CCICED), 2015]。在欧洲经济区和西巴尔干合作国家(EEA-39)还有大约300万处可能受到潜在污染的地方(EEA, 2014),美国有超过1 300处已污染或潜在污染的地点被列入超级基金国家优先事项清单[美国环境保护署(US EPA), 2013]。整个澳大利亚被污染的地点总数估计为8万处(DECA, 2010)。虽然这些数字帮助我们在了解某些活动对土壤的影响方面提供了信息,但它们不能反映世界各地土壤污染的全部情况,而且它们也显示出了现有信息的不足以及不同地理区域在登记污染站点方面上的差异(Panagiotakis和Dermatas, 2015)。在低收入和中等收入国家,由于这些数据和信息的缺乏,使得世界上最大的全球性问题之一——土壤污染成为国际社会中无形的问题。综上所述,对土壤污染进行全球评估的任务十分迫切。

幸运的是,世界各地对土壤污染的重要性的认识正在增强,因此对土壤污染评估和修复的研究也在增加(图1)。经修订的《世界土壤宪章》(FAO, 2015b)建议各国政府执行关于土壤污染的条例,限制污染物积累水平,从而保障人类的健康和生活。各国政府还被敦促协助修复超过规定水平的污染土壤,从而保护人类健康和环境。在第五届全球土壤伙伴关系大会上土壤污染占据了中心议题位置[全球土壤伙伴计划(GSP), 2017]。最近,联合国环境大会(UNEA-3)通过了一项呼吁在可持续发展框架内,加快处理和管理土壤污染方面行动和合作的决议。170多个国家达成的这一共识清楚地表明,土壤污染已具有全球相关性,这些国家具有为解决污染问题制定具体解决方案

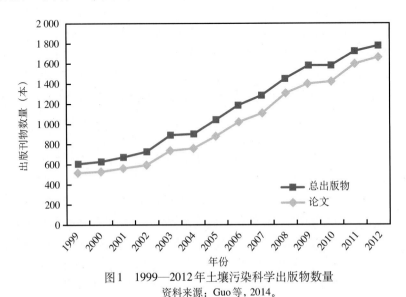

图1　1999—2012年土壤污染科学出版物数量
资料来源:Guo等, 2014。

的意愿[联合国环境规划署（UNEP），2018]。在国家级别上，世界上许多国家都通过国家条例来保护土壤，防止土壤污染，并处理历史土壤污染问题。2017年下半年，在爱沙尼亚担任欧盟理事会主席期间，土壤成为欧盟成员讨论的主要话题之一，并聚焦于土壤在食品生产中发挥的关键作用。在中国，过去几年里人们对土壤污染的担忧有所增加，部分原因是由于此问题与人类健康直接相关。其他发展中国家最近也通过了相关规定，以防止和控制土壤污染，并确定土壤质量[巴西国家环境委员会（Conselho Nacional do MeioAmbiente），2009；秘鲁环境部（MINAM），2017；智利国家环境委员会（MMA），2013]。

"soil contamination"经常作为"soil pollution"的同义词被用作土壤污染。全球土壤伙伴关系（GSP）下的政府间土壤技术小组（ITPS）正式定义了这两个术语（FAO和ITPS，2015）。"soil contamination"适用于当一种化学品或物质的浓度高于自然中本该有的浓度，但不一定会造成危害的情况；而"soil pollution"是指化学品或物质出现位置不当或土壤中浓度高于正常浓度，且对非靶向生物体会产生不利影响的情况。

当前面临的一个重要问题是"正常浓度"很难被准确定义。建立土壤中原本不存在的人为源物质的危险浓度比较容易，但对重金属和类金属，要确定这样的浓度则可能是一项挑战，因为重金属和金属可以来自岩石和矿物的风化作用。在这种情况下，在建立阈值之前需要考虑母质、气候和风化速度。此外，土地使用及管理操作会影响土壤中存在物质的背景浓度。在提及建议浓度时，各国和各区域之间也存在许多差异，不仅仅是关于这个值本身，同时还关于定义它的一些名称，包括筛选值、阈值、可接受浓度、目标值、干预值、清除值以及许多其他的名称（Beyer，1990；Carlon等，2007；Jennings，2013）。因此，对全球土壤污染的实际状况进行全球性的研究并进行比较是极其复杂的。这是对区域和全球土壤污染进行评估时的主要挑战之一。

科学家之间就概念和定义达成一致，将有助于决策者和利益相关者确定世界不同地区用于评估和处理土壤污染的相关策略和技术。使用一种共同的、简化的语言也将有助于更好地理解土壤污染问题。

1.2 点源和土壤污染扩散

如前所述，土壤污染可由有意和无意的活动导致。这些活动包括污染物直接沉降到土壤中，以及通过水或大气沉降间接导致土壤污染的复杂环境过程（Tarazona，2014）。在下面的章节中将介绍不同类型的土壤污染。

1.2.1 点源污染

土壤污染可由某一特定区域内某一特定事件或一系列事件引起，污染物在这些事件中被释放到土壤中，污染的来源和特征很容易识别，这种类型的污染称为点源污染。人类活动是点源污染的主要来源，例如旧的工厂、未充分处理的废物和废水、管控不严格的垃圾填埋厂、过度使用农用化学品、多种类型的泄露以及其他很多情况。在世界上的许多地区，使用粗劣的环境标准进行的如采矿和冶炼等活动也是含重金属污染物污染的来源（Lu 等，2015；Mackay 等，2013；Podolský 等，2015；Strzebońska，Jarosz-Krzemińska 和 Adamiec，2017）。点源污染的其他例子还有与石油产品有关的芳香烃和有毒金属。在这方面，相关的事件主要有格陵兰的储罐设施泄漏事件，造成了芳香烃和有毒金属水平超过丹麦环境质量标准（Fritt-Rasmussen 等，2012），以及德黑兰炼油厂储罐的意外泄漏事件等（Bayat 等，2016）。

点源污染在城市中非常普遍。道路附近的土壤含有大量的重金属、多环芳烃和其他污染物（Kim 等，2017；Kumar 和 Kothiyal，2016；Venuti, Alfonsi 和 Cavallo，2016；Zhang 等，2015b）。旧的或非法的垃圾填埋场里，未得到恰当处置或未根据其毒性进行处置的废物（如电池或放射性废料），以及处置不当的污水污泥和废水，也是重要的点源污染源（Baderna 等，2011；Bauman-Kaszubska 和 Sikorski，2009；Swati 等，2014）。最后，由工业活动引起的点源污染会对人类健康构成威胁。例如，中国有超过5 000个棕色地带正在影响其居民的健康（Yang 等，2014）。位于城市中心的城市棕色地带，曾经是工业活动的场所，后来被重新安置。

1.2.2 扩散污染

扩散污染是一种分布范围广，可以在土壤中积累，污染来源复杂且不容易识别污染源的污染。扩散污染是指其他介质中的污染物经过排放、转化和稀释然后转移到土壤的过程（FAO 和 ITPS，2015）。扩散污染包括污染物通过空气-土壤-水体系统的传输。所以，为了充分评估这类污染，需要对这三个过程进行复杂的分析（Geissen 等，2015）。因此，扩散污染难以分析，且跟踪和界定其空间范围也具有挑战性。许多造成局地污染的污染物在环境中的变化过程还没有被完全了解，他们在造成局地污染的同时也可能涉及扩散污染（Grathwohl 和 Halm，2003）。扩散污染的例子有很多，包括核电站以及战争活动造成的污染；未经控制的废物处置和在集水区及其附近排放的污水；污水污泥在土地上的应用；农业上使用的含重金属的农药和化肥、持久性有机污染物；过量的营养物质和由地表径流向下游输送的农药；洪水事件；大气传输和沉降；水土流失（图2）。虽然扩散污染的严重程度和范围尚不清楚，但是其对环境和人类健康有严重影响。

已有大量证据证明，土壤上层富含许多金属和其他元素，这些与自然源和人为源的大气沉降有关（Blaser 等，2000；Steinnes 等，1997；Steinnes, Berg 和 Uggerud，2011）。在北半球的几乎每一寸土壤都含有比本底水平高的放射性核素，即使在北美和亚洲东部的偏远地区也是如此。由于切尔诺贝利灾难性事故产生的核沉降物，其放射性核素将在土壤中存在几个世纪（Fesenko 等，2007）。在距离切尔诺贝利200千米的地区，放射性核素，比如 $^{239/240}$Pu 或 ^{241}Am，浓度减少50%需要50年以上的时间。

由于这些不同污染源的污染类型多种多样，我们需要在科学和技术方面进一步加大投入，来开发新的测量和监测方法，以期更好地了解大气沉降过程和扩散污染程度。

图2　农药在环境中的运输途径
资料来源：联合国粮农组织，2000年

1.3 土壤污染来源

1.3.1 自然、地球成因来源

在通过环境立法对环境问题规定干预限度时，区分背景值和基线值对定义污染程度至关重要（Albanese 等，2007）。背景值表示地质自然含量，而基线值表示某一元素在某一地点的表层土壤环境中的实际含量（Reimann, Filzmoser 和 Garrett, 2005；Salminen 和 Gregorauskiene, 2000）。

一个地区土壤的背景浓度与该地区的地质化学特性和导致土壤形成的环境条件密切相关。因此，采用平均或全球间隔划分的方法不能用来确定区域层面上的背景水平（Horckmans 等，2005；Paye, Mello 和 Melo, 2012）。例如，在母岩类型中微量金属浓度自然变化的影响下，土壤中重金属的变化可以超过两到三个数量级（Shacklette 和 Boerngen, 1984）。

一些土壤母质是某些重金属和其他元素（如放射性核素）的自然来源，而这些元素在浓度升高时可能对环境和人类健康构成威胁。砷（As）污染是世界各地主要环境问题之一。砷的自然源包括火山释放物（Albanese 等，2007）以及含砷矿物和矿石的风化（Díez 等，2009），也包括由含硫化物的岩石风化自然形成的砷黄铁矿矿化带（铁帽）（Scott, Ashley 和 Lawie, 2001）。这些矿物有许多具有高度的空间变异性，往往在较深的地层中具有较高浓度（Li 等，2017）。与此同时，自然源的As与人为来源的As相比，其生物可利用性较弱（Juhasz 等，2007）。

土壤和岩石也是放射性气体氡（Rn）的自然源。这些气体氡在土壤结构及其孔隙的影响下从土壤深层扩散到地表（Hafez 和 Awad, 2016）。天然的高放射性物质在酸性岩浆岩中很常见，如富长石岩和富伊利石岩（Blume 等，2016）。Gregorič 等人发现与其他任何土壤或岩石类的土壤相比，含碳酸盐的土壤会释放出更高的氡（Gregorič 等，2013）。岩石和土壤中其他天然放射性核素的参考数据见表1。

表1 岩石和土壤中天然放射性核素的比活度　　　　单位：贝可/千克

岩石/土壤	^{40}K	^{226}Ra	^{232}Th
沙岩	461	35	4
黏土岩	876	无	41
片岩（法兰克尼亚）	1 000	3 000	60
碳酸盐岩	97	<10	5

(续)

岩石/土壤	^{40}K	^{226}Ra	^{232}Th
酸性火成岩	997	37	52
基性火成岩	187	10	8
黄土发展而成的土壤	无	41	54
花岗岩发展而成的土壤	<1 100	65～75	38～72
石英岩发展而成的土壤	<300	54～56	63～70
千枚岩发展而成的土壤	无	40～70	50～80

资料来源：Blume等，2016。

火山爆发或森林火灾等自然事件也会造成自然污染，因为过程中许多有毒元素会被释放到环境中。这些有毒元素包括二噁英类化合物（Deardorff，Karch和Holm，2008）和多环芳烃（PAHs）。已经在留尼汪岛的火山土壤中发现高浓度的重金属，其中一些重金属（主要是汞）与活跃的火山活动有关，高浓度的铬（Cr）、铜（Cu）、镍（Ni）和锌（Zn）则与母质的风化过程有关（Dœlsch，Saint Macary和Van de Kerchove，2006）。在印度尼西亚的火山土壤中也有高浓度的铬和镍，这与土壤地球化学成因有关（Anda，2012）。然而，由于植物的再生能力和适应能力，使得这种自然污染通常不会造成环境问题（Kim，Choi和Chang，2011）。但当生态系统受到的外部压力超过其恢复力和反应能力时，环境问题就出现了。

多环芳烃也可以在土壤中自然产生。它们通常是宇宙起源之初就存在的，在宇宙尘埃样品和陨石中比较常见（Basile，Middleditch和Oro，1984；Li，2009），也有一部分源于土壤有机质中蜡质的成岩蚀变过程（Trendel等，1989）。通常还原条件更有利于多环芳烃生物生成过程的进行（Thiele和Brummer，2002）。

天然存在的石棉是一种纤维矿物，自然存在于由超镁铁质岩石形成的土壤中，特别是蛇纹石和角闪石。与石棉有关的主要风险是与人类提取活动有关的吸入暴露，而它在土壤中的自然存在对环境造成的风险可以忽略不计。但是，石棉很容易通过风蚀而传播，其移动将取决于石棉材料的特性、土壤特性、湿度和天气状况（Swartjes和Tromp，2008）。因为石棉是一种致癌物质，吸入后对人体健康会造成高风险危害，当石棉从城市地区附近的土壤中释放出来时，就会引起环境问题（Lee等，2008）。

1.3.2 人为源

几个世纪的人为活动已经在世界各地造成了普遍的土壤污染问题（Bundschuh 等，2012；DEA，2010；EEA，2014；FAO 和 ITPS，2015；Luo 等，2009；SSR，2010）。

土壤污染的主要人为源是在工业活动中使用或作为副产品产生的化学品、家庭和城市废物（包括废水、农药和石油衍生产品）（图3）。这些化学物质会被无意（如石油泄漏或垃圾填埋场的沥滤）或有意地释放到环境中（如使用化肥和农药、未经处理的废水灌溉或污水污泥利用）。

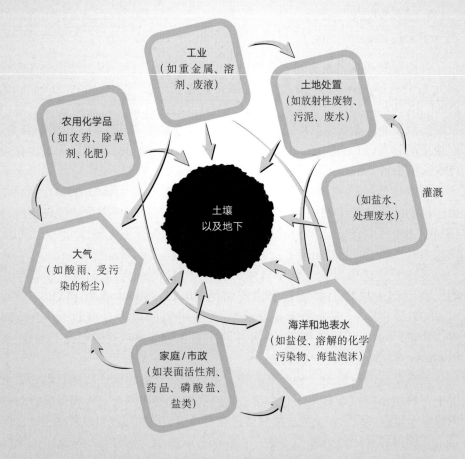

图3 土壤-地下化学污染的潜在相互关联途径
资料来源：Yaron, Dror 和 Berkowitz, 2012。

1.3.2.1 工业活动

工业活动中使用的化学品种类繁多，对环境的影响也很大。

工业活动会向大气、水和土壤释放多种污染物。气态污染物和放射性核素释放到大气中，可通过酸雨或大气沉降直接进入土壤；以前的工业用地可能因错误的化学储存或直接将废物排入土壤而使土壤受到污染；用于热电厂和许多其他工业过程的冷却水和其他液体可以排入河流、湖泊和海洋，造成热污染，并且其携带的重金属和氯会影响水生生物和其他水体。另外，人类活动可能通过沙尘、原材料泄漏、废水、终端产品、燃料灰以及火灾等在工业场所构成影响（Alloway，2013）。

根据欧洲关于综合污染的预防和控制指令（EC，1996年），可能造成污染的活动可分为六大类：①能源产业；②金属生产和加工业；③矿产行业；④化工和化工设施；⑤废物处理；⑥其他活动（包括造纸和纸板生产、纤维或纺织品制造、皮革鞣制、屠宰场、家禽密集饲养或养猪、使用有机溶剂的设施以及碳或石墨的生产）（García-Pérez等，2007）。

盐碱化是全球土壤的另一个主要威胁。靠近某些工业活动的许多土壤会受到影响，主要是那些与氯碱、纺织品、玻璃、橡胶生产、兽皮加工和皮革制革、金属加工、医药、石油和天然气钻井、颜料制造、陶瓷制造、肥皂和洗涤剂生产有关的工业活动（Saha等，2017）。

1.3.2.2 采矿业

采矿自古以来就对土壤、水和生物群产生重大影响（FAO和ITPS，2015）。在世界各地，可以发现许多与采矿活动相关的严重土壤污染实例（Alloway，2013）。

为了分离矿物而进行的金属冶炼已经向土壤中释放了许多污染物。采矿和冶炼设施会向环境中排放大量重金属和其他有毒元素，而它们会在活动结束很长一段时间后仍持续存在（Ogundele等，2017）。

有毒的矿采废料主要由细颗粒物构成，其中含有不同浓度的重金属，会堆积在尾渣中。这些含污染物的颗粒通过风蚀和水侵蚀而传播，有的会到达农业土壤。例如，Mileusnić等人发现纳米比亚尾矿坝附近的农田中铅和铜含量很高（Mileusnić等，2014）。在印度一个废弃的铬石棉矿废料附近的农业土壤中以及在这些土壤种植的作物中也发现了有毒的铬和镍，这些对人类和牲畜健康会构成高风险（Kumar和Maiti，2015）。

使用富含天然放射性物质的磷矿生产肥料时会产生一种被称为磷石膏的副产品，由于其中含有 ^{238}U 衰变产物，如镭（^{226}Ra）和钋（^{210}Po），磷石膏会保持近80%的原始放射性。这些工业产生的放射性污染源会对周围的生态系统和生物体构成威胁（Bolívar，García-Tenorio和García-León，1995）。

在石油和天然气的提取过程中，由于原油和卤水的泄漏，会造成严重的点源土壤污染。卤水的含盐量很高，还可能含有有毒的微量元素和天然放射性物质。卤水泄漏事件是普遍存在的，例如，Lauer等人指出，自2007年以来，在北达科他州的巴肯地区，大约有3 900起与非常规油气生产（包括水力压裂）有关的卤水泄漏事件（Lauer, Harkness 和 Vengosh, 2016）。油井和管道泄漏的原油也是产油地区土壤污染的主要来源。

1.3.2.3 城市与基础运输设施

住房、公路和铁路等基础设施的发展大大加快了环境的退化。其对土壤的不利影响主要表现为土壤封闭和土地消耗。除了这些已知的威胁之外，基础设施活动的另一个主要影响是多种污染物会进入土壤系统。尽管基础设施活动对土壤污染是一个重大威胁，但在土地规划和影响评估方面对它的考虑却非常少。

城市中心及其周围与交通相关的活动也是构成土壤污染的主要来源之一，因为内燃机的排放物会通过大气沉降和汽油泄漏到达100米以外的土壤，同时也源于这些活动自身以及它们作为一个整体所引起的变化（Mirsal, 2008）。如果交通设施的排水系统没有得到良好的维护，在降雨期间以及径流中由交通产生的飞溅泡沫可能对环境污染造成严重影响，因为这些飞溅泡沫可能会导致富含重金属的颗粒发生转移，这些颗粒来自车辆金属部件的腐蚀和轮胎与路面的磨损（Venuti, Alfonsi 和 Cavallo, 2016；Zhang 等, 2015b），并且可能含有其他污染物如多环芳烃、橡胶和塑料衍生化合物（Kumar 和 Kothiyal, 2016；Wawer 等, 2015）。在城市和城市周边土壤中与道路和高速公路有关的土壤污染尤为严重，尤其是当邻近地区有粮食生产时，它就可能成为主要威胁。污染在路边土壤中发生的主要过程是落叶、根系吸收和生物可利用重金属向地上组织转移（Hashim 等, 2017；Kim 等, 2017；Zhang 等, 2015b）。此外，在路边的放牧活动也很常见，摄入受污染的土壤和植物，会构成潜在的污染物的膳食转移，从而影响动物和人类健康（Cruz 等, 2014）。

与运输有关的土壤污染的另一个重要来源是含铅汽油对土壤的铅污染。Mielke 和 Reagan 的研究报告指出，超过1 000万吨的铅可以通过机动车行驶过程转移到全球环境中，仅在美国就有590万吨（Mielke 和 Reagan, 1998）。由此造成的土壤污染集中在道路周围，且在核心城区尤其严重。

堆填区处置都市废物，不论是否违法，以及未经处理的废水，都是土壤中重金属、难生物降解的有机物以及其他污染物的重要来源。在大多数发达国家，废物、固体和液体的处置和回收都有严格的规定[欧盟委员会（EC），1986；美国联邦注册（US Federal Register）, 1993]，但有些国家的废物处理和处置仍然对环境和人类健康存在威胁。

许多家用化学品，特别是大批量使用的化学品，如洗涤剂和个人护理用品，最终也会成为生活污水成分。市政污水处理生成的有机污泥是许多个人护理用品的主要汇集，而这些有机污泥在土地上的应用可能会将这些污染物引入到陆地和水生环境中。与此同时，长时间使用滴滴涕控制疟疾等病媒传播疾病也导致了城市和城郊地区的土壤污染（Mansouri等，2017）。

含铅涂料是城市地区铅污染的主要来源。建筑翻新或拆除过程中含铅涂料被粉碎成粉尘或小颗粒后进入环境，从而污染土壤（Mielke和Reagan，1998）。在美国，在1929年到1989年，铅主要被用于含铅汽油，在1884年到1989年主要用于白铅喷绘颜料，在1920年到1929年，含铅涂料的使用达到高峰（Mielke和Reagan, 1998）。

塑料也是污染的一个主要来源。塑料被广泛用于食品包装、购物袋和家居用品（如牙刷、笔、洁面膏和许多其他常见物品）。塑料对于全球环境具有重要影响。一般而言，它们在环境中长久存在，而且可以在海洋和垃圾填埋场中广泛累积，同时生产工厂所在地的土壤里也大量存在。聚合物的生化性质通常被认为是惰性的，不会对环境构成威胁。然而，由于聚合反应很少能充分反应，所以在塑料材料中可以发现未反应的单体或低聚物（Araújo等，2002）。最危险的单体是聚氨酯、聚丙烯腈、聚氯乙烯、环氧树脂和苯乙烯共聚物，它们被认为具有致癌性或同时具有致癌性和诱变性（Lithner, Larsson和Dave, 2011）。此外，数千种不同的添加剂，如溴化阻燃剂、邻苯二甲酸盐和铅化合物被用于塑料的生产。这些添加剂中有许多被认为是有害的，会对内分泌功能有破坏作用，对生物体有致癌和诱变作用（Darnerud, 2003；Heudorf, Mersch-Sundermann和Angerer, 2007；Lithner, Larsson和Dave, 2011）。所有塑料，从宏观到纳米尺度，都面临被淋溶以及吸附持久性有机污染物和多环芳烃等有害物质的风险（Björnsdotter, 2015）。与此同时，它们还会大量积累重金属（Mato等，2001）。体积和表面积是影响淋溶和吸附行为的重要因素，通常而言，颗粒越小，表面积与体积比越大。因此，小颗粒释放或结合化合物的能力比大颗粒强。

塑料可以通过污水处理厂进入土壤和水生系统，也可以通过风从垃圾填埋场被转移而悬浮至空气中，然后在空气中广泛传播。在常用塑料覆盖的农田中，土壤中会含有大量的塑料。塑料在水生生物和生态系统中的存在和影响已有很充分的研究记录（Browne等，2008；Thompson, 2004）；然而，仍然需要就使用塑料聚合物及其产品对人类健康和陆地生态系统的风险进行评估（Lithner, Larsson和Dave, 2011；Rillig, 2012；Rocha-Santos和Duarte, 2015）。到目前为止几乎没有人对塑料在土壤中的行为过程进行过相关的研究。

1.3.2.4 废物和污水的产生和处理

随着全球人口的增加，垃圾的产生量也在增加。在发展中国家和不发达的国家，高人口增长率和不断增加的废水和污泥产量，再加上缺乏废物处理的市政管理，导致了环境污染的危险局面。据世界银行的一份报告（Hoornweg 和 Bhada-Tata, 2012）显示，在2012年，城市固体废弃物的全球产量约为13亿吨，从撒哈拉以南非洲的人均每天0.45千克到经济合作与发展组织国家的人均2.2千克不等。然而，未来更令人担忧，预计到2025年废物产生量将增加到22亿吨。

垃圾填埋和焚烧是垃圾处理最常见的两种方式。在这两种情况下，许多污染物，如重金属、多环芳烃、药物化合物、个人护理产品及其衍生物在土壤中积累（Swati 等，2014），累积的污染物要么是直接来自污染土壤和地下水的垃圾渗滤液，要么是来自焚烧厂的飞灰沉降物（Mirsal, 2008）。Baderna 等人在垃圾渗滤液中发现了一种复杂的混合污染物，可以改变地下水质量，进而影响食物链（Baderna 等，2011）。

回收铅电池的企业已被认定为世界各地土壤污染的主要来源，尤其是在非洲。那里的铅电池行业在过去几年中显著扩张，并将继续增长，但那里的监管薄弱甚至缺乏（Gottesfeld 等，2018）。铅电池工业和回收工厂靠近社区，对人类健康构成高风险，这一结论可以从铅含量超过了筛选标准的血液样本中得出 [美国有毒物质和疾病登记处（US Agency for Toxic Subtances and Disease Registry），2011；Zahran 等，2013]。

21世纪带来了通信的进步和重要技术的发展。电气和电子设备的生产量正在全世界范围内迅速增长，并将持续增长，发展中国家将在未来十年成为主要的生产国（Robinson, 2009）。然而，一旦设备过时或不再发挥作用，它们最终会变成废物。电子垃圾中包含有价值的元素，如铜和金，但也有许多其他有害物质，使其不可能像处理普通城市垃圾一样被处理。在欧洲和北美洲，大部分电子垃圾仍然无法回收（Barba-Gutiérrez, Adenso-Díaz 和 Hopp, 2008；Sthiannopkao 和 Wong, 2013），然而电子垃圾已成为发展中国家或新兴工业化国家的收入来源之一。据Itai 等人报道，在加纳的一个电子垃圾回收点，重金属和稀有金属（In、Sb、Bi）浓度很高，这表明应该在风险评估办法中纳入这些金属（Itai 等，2014）。然而，正规的回收中心只占回收行业的25%，电子垃圾大多是在非正规部门回收的，这些部门通常使用的是较为原始的技术，比如燃烧电缆来收集铜。这些技术会释放大量有害物质（阻燃剂、二恶英类化合物、多环芳烃、重金属），他们并没有考虑和采取对环境或人类健康的保护措施（Perkins 等，2014）。

利用污水污泥改良土壤可能是有益的，因为它向土壤中添加了有机质和养分。然而，如果污水污泥在使用前没有进行预处理，许多污染物如重金属就会在土壤中累积，最终进入食物链。虽然在欧洲污水污泥的使用是受到管制的，但是并不代表在所有地方都受管制。

在干旱和半干旱地区，将处理过的废水用于农业灌溉是解决缺水问题的常见方法（Jefatura del Estado，2001；Keraita 和 Drechsel，2004；Uzen，2016）。例如，在以色列超过80%的城市污水被再利用（Katz，2016）；巴基斯坦有26%的蔬菜生产是用废水灌溉的（Ensink 等，2004）；西班牙的干旱地区使用再生废水解决了缺水问题，同时这也是增加土壤营养和提高作物产量的一种方式（Dorta-Santos 等，2014）。然而，在没有水质准则和相关法规的国家，废水的使用可能是一个问题。如果污染物质在废水处理后没有被完全清除，或者废水未经处理，则废水的使用可能会导致重金属、盐类、个人护理用品和病原体的沉积（Dalkmann 等，2014；Flores-Magdaleno 等，2011；Pedrero 等，2010）。

1.3.2.5 军事活动和战争

直到20世纪，大多数冲突都是地方性的，对土壤的影响相对较小。然而，现代化战争使用的是不可降解的、具有毁灭性的武器，以及在冲突结束后仍然可以在受影响的土壤中存留数个世纪的化学物质（FAO 和 ITPS，2015）。在战争时期以及和平时期，诸如试射之类的军事活动，都会极大地改变土壤性质。这些土壤的完全恢复或者部分恢复需要很多年，有时甚至需要几个世纪（Certini，Scalenghe 和 Woods，2013）。

第一次和第二次世界大战给欧洲造成了严重的污染遗留问题（地雷、弹药和化学品残留物、放射性和生物毒性试剂），这些不仅存在于战场上，在射击区、营房和武器仓库等地点也都存在。这些遗留使得这些地区的一些土壤不适合任何类型的开发。从20世纪初至今，大约有1.1亿枚地雷和其他未爆弹药散布在各大洲的64个国家（Kobayashi，2012）。

军需品的处置，以及由于情况紧急而造成的生产过程中的疏忽，可能会对土壤造成长期污染。这类污染几乎没有公开的证据，这很大程度上是由于许多国家政府对发表与战争有关的材料设置了限制。例如，在柏林，有一千多公顷的土地受到严重污染（Schafer,1995）；在苏格兰西部的格鲁纳德岛，尽管采取了补救措施，但仍被用作生物武器的炭疽孢子污染（Szasz,1995；WHO,2008）。第二次世界大战期间储存的芥子气也污染了一些地点长达50年之久（Watson 和 Griffin,1992）。

1.3.2.6 农业及畜牧业活动

土壤污染物的不同农业来源包括农业化学来源，如化肥、动物粪肥、各类农药（图4）。这些农用化学品中的微量金属，如铜（Cu）、镉（Cd）、铅

土壤污染：一个隐藏的现实

（Pb）和汞（Hg），也被认为是土壤污染物，因为它们会影响植物的新陈代谢并降低作物产量。灌溉水源中如果包含有废水和城市污水，也会造成土壤污染。过量的氮和重金属不仅是土壤污染的一个来源，而且当它们进入食物链后，会对粮食安全、水质和人类健康构成威胁（FAO 和 ITPS，2015）。

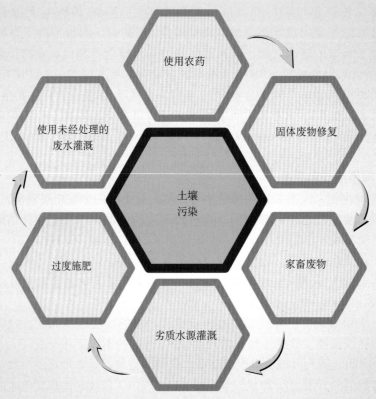

图4　土壤污染的农业源

农业活动产生的点源污染包括用于机器燃料的碳氢化合物在农业土壤中的意外泄漏或农业化学品在运输和储存阶段的意外泄漏。

如前所述，畜牧业生产也是土壤污染的来源之一，尤其是当废物没有得到适当的管理和处置时：尿液和粪便中含有的寄生虫以及药物可以在土壤中持续累积（Zhang 等，2015a）。许多药物是脂溶性的，不易降解，所以有可能被保留在污泥或粪肥中，而这些污泥或粪肥又反过来被用作肥料，这种保留会极大地影响土壤中的微生物和其他有益生物体（Halling-Sørensen 等，1998）。

化肥和有机粪肥的过度施用，或者肥料中主要营养物质（氮和磷）的使用效率低下是造成与农业相关的环境问题的主要原因（Kanter，2018）。氮和磷这两种营养物质是扩散污染的来源之一。过量的氮会通过温室气体排放而流失到大气中，过量的磷会导致邻近水源的富营养化。化肥的过度使用会导致土壤盐碱化、重金属积累、水体富营养化和硝酸盐积累，这些都会成为环境污染源，同时也会对人类健康构成威胁。化肥工业还被认为是Hg、Cd、As、Pb、Cu、Ni和Cu等重金属以及^{238}U、^{232}Th和^{210}Po等天然放射性核素的来源。因此，正确处理和管理化肥是避免土壤污染的关键（Stewart等，2005）。

堆肥和牲畜排泄物是营养物质的重要来源。它们有助于实现循环经济，减少废物对环境的影响，增加土壤中的有机质和氮含量，同时减少农业生态系统的外部投入（Shiralipour, McConnell 和 Smith,1992）。Xia等人的报告指出，当用粪肥部分替代合成化肥时，作物产量总体可以增加4.4%（Xia等，2017）。该替代部分提高了作物对氮和其他养分的吸收，而且显著减少了挥发、侵蚀和淋溶等造成的氮损失，这主要是由于其养分释放缓慢并促进了微生物对生物有效态氮的固定。然而，农作物产量的增加取决于粪肥和作物类型（Wang等，2016；Xia等，2017）。此外，堆肥和粪肥是有机质的重要来源（Zhao等，2014b）。

联合国粮农组织生产和植物保护司最近的一份报告显示，从1961年到2016年，全球所有牲畜的粪肥生产量增长了66%，从7 300万吨氮增至1.24亿吨氮，施入土壤的粪肥从1 800万吨氮增加到2 800万吨氮，同时牧场粪肥施用量从4 800万吨氮增加到8 600万吨氮（Raffa等，2018）。

尽管它们对农业有潜在效益，但充分的科学证据表明土壤中重金属含量、病原体和兽药抗生素残留物都有所增加，这可能导致具有抗生素耐药性的细菌在经过粪肥改良过的土壤中增殖。牲畜粪便中的重金属主要来源于饲料（Nicholson等，1999），而抗生素被用于预防和治疗疾病以及作为促生长剂使用（Kumar等，2005）。为了确定重金属的主要来源，Nicholson等人在英格兰和威尔士的农田中进行了一次调查（Nicholson等，2003）。除了大气沉降这一主要污染源，牲畜粪便和污泥也被认定是土壤污染的重要污染源。这些对土壤中的Zn、Cu、Ni、Pb和Cr污染有重要影响（Nicholson等，1999；Wang等，2016）。

农药是指旨在防止、消灭或控制任何害虫引起的危害或者用于食品、农产品、木材和木制品生产、加工、储存、运输或销售的一类药物或药物混合物（FAO, 2006）。杀虫剂是所有农药的一个子集。自第二次世界大战以来，当DDT的杀虫特性被发现的时候，它们就被更大规模地使用到了环境中。农药的使用为日益增长的人口提供了食物；然而，过度使用农药会对人类健康和环境产生负面影响（Popp, Pető 和 Nagy, 2013；FAO 和 ITPS, 2017）。农药对土壤生物的负面影响已得到了广泛研究（Bünemann, Schwenke 和 Van Zwieten, 2006；

Jacobsen 和 Hjelmsø，2014；Komárek 等，2010；Nguyen 等，2016；Ockleford 等，2017；Puglisi，2012），且有些健康问题已经证实与接触农药和其他农用化学品有关。农药对人类健康的主要威胁是终生暴露在低剂量环境下（WHO，1993），并且这种接触的直接短期影响不明显。

考虑到在危险化学品方面需要进行统筹协调，《鹿特丹公约》（以下简称《公约》）于1998年9月10日获得通过。它旨在促进各缔约方在危险化学品和农药的国际贸易中共担责任并努力合作，以保护人类健康和环境免受潜在损害。《公约》通过促进关于这些化学品特性的信息交流，制定关于这些化学品进出口的国家决策流程，并将这些决策宣传给各缔约国，从而有助于促进在无害环境下使用这些化学品。

《公约》的好处是可以通过具有法律约束力的事先知情同意程序防止化学品的不必要贸易。它使成员国政府能够通过交换关于被禁止或严格限制的化学品的信息，互相提醒潜在的危险，并就这些化学品做出正确的决定。它通过出口通报和规定以及鼓励对出口化学品使用统一标签，使危险化学品的国际贸易更透明，防止其泛滥。《公约》还要求提供技术援助，以帮助建立安全管理化学品所需的基础设施和能力。

《农药管理国际行为守则》（FAO，2003）为农药整个生命周期的管理提供了推荐性框架和标准。该守则主要面向政府当局和农药行业，但也与其他利益相关者有关。该守则由技术指南和工具包支撑，例如登记工具包（http://www.fao.org/pesticide-registration-toolkit/en/）以及废弃农药环境管理工具包。

过期、不需要和被禁止的农药库存不断地对它们所在区域的人类健康、环境和可持续发展构成威胁。过期农药积累的原因通常是有据可查的。它们包括储存和管理不善的农药、国际禁令中禁止使用的有害农药、不恰当捐赠的农药、订购和供应过多以及采购用于控制迁移害虫防治的战略库存农药，而这些农药后来没有完全使用或没有必要使用。农药通常储存在非常恶劣的条件下，导致集装箱变质并泄漏到周围环境中并最终影响土壤和地下水质量。联合国粮农组织废弃农药环境管理工具包的第1～4卷，旨在协助各国对过期农药库存进行风险管理。第6卷旨在提供切实可行的方法，协助各国制定战略，管理被农药污染的土地。由于能否降低风险很大程度上取决于现场调查和相关风险评估的准确性，因此该卷将与第5卷紧密结合使用，其可对农药污染土地的环境管理计划设计起到重要作用。

农业污染的其他来源包括集中饲养动物造成的砷污染、塑料覆盖导致的塑料残留，使用受污染的地下水进行灌溉，以及很多其他方面导致的污染。Liu和Cang等研究了由家禽和家畜造成的土壤污染，他们都发现家禽和家畜饲养业会导致大量的重金属污染（Cang等，2004；Liu等，2015）。

1.4 土壤主要污染物

如前所述，向环境排放污染物通常源自人为过程。虽然某些元素和化合物在土壤中是天然存在的，人类的干预仍是导致土壤污染的主要驱动因素。以下各节只讨论影响农业区域的一小部分最常见的污染物，以及这些化合物作为污染物的特性。相关研究根据不同化学特征已经对污染物进行了分类，不同类别有重叠之处。Swartjes提出了一种污染物的系统分类（图5），这对于更好地理解它们有一定帮助（Swartjes, 2011）。

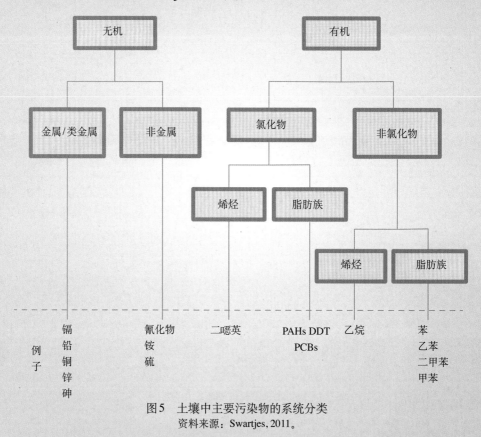

图5　土壤中主要污染物的系统分类
资料来源：Swartjes, 2011。

1.4.1 重金属和类金属

"重金属"是指原子质量相对较高（>4.5克/立方厘米）的金属和类金属，如铅（Pb）、镉（Cd）、铜（Cu）、汞（Hg）、锡（Sn）和锌（Zn）等可引起毒性问题。通常与重金属一起考虑的其他非金属包括砷（As）、锑（Sb）和硒

（Se）（Kemp，1998）。这些元素在土壤中的天然浓度比较低。它们中有许多是植物、动物和人类所必需的微量元素，但是高浓度可能造成植物毒性并损害人类健康，因为它们具有不可生物降解的性质，容易在组织和生物体中积累。

重金属的主要人为来源是工业区、矿山尾矿、重金属废物的处置、含铅汽油和油漆、化肥的应用、动物粪肥、污水污泥、农药、废水灌溉、煤炭燃烧残渣、石化化工产品的泄漏和各种来源的大气沉降（Alloway，2013）。

重金属是自然界中难治理的一类最持久、最复杂的污染物。它们不仅会降低大气、水体和粮食作物的质量，而且还会威胁动物和人类的健康和福祉。因为金属与大多数有机化合物不同，它们不会被代谢分解，所以它们会在生物体组织中积累。在重金属中，Zn、Ni、Co、Cu对植物的毒性相对较大，As、Cd、Pb、Cr、Hg对高等动物的毒性相对较大（McBride，1994）。

就食物链污染而言，需要考虑的最重要元素是As、Cd、Hg、Pb和Se（McLaughlin，Parker 和 Clarke，1999）。土壤中As的主要来源是农药化合物、采矿和冶炼活动，同时它们也可以来自食用含有As添加剂饲料的牲畜的粪便。一些母质材料富含As，因此它们的风化作用也可能是高浓度As的一个来源。

铜基的无机和有机农药中的微量金属是环境和毒理学方面研究的一个主要问题（Komárek 等，2010）。铜很容易被土壤中有机质以及铁和锰的氧化物固定，所以可以较高浓度地保留在土壤上层。研究者在土壤的潜在可用部分中发现了大量来源于杀菌剂的铜（Pietrzak 和 McPhail，2004）。

1.4.2 氮和磷

氮是蛋白质、DNA、RNA、激素、酶和维生素等所有生命结构的重要组成部分。它可以以有机和无机两种形式存在，同时也有许多不同的氧化态。它的可用形态根据特定生物的需求不同而不同。它的不活泼形态——氮气（N_2），可以通过微生物活动被吸收固定。植物则需要更多的化学可用形态，例如铵盐（NH_4^+）和硝酸盐（NO_3^-），而动物需要的形态复杂，如氨基酸和核酸（Yaron，Dror 和 Berkowitz，2012）。

磷是所有生物的主要微量营养元素之一。它是组成DNA和RNA等生物分子的元素之一，同时它也可形成三磷酸腺苷来转运细胞能量。

为了养活不断增长的人口，并满足世界各地许多贫瘠土壤的养分需求，在整个20世纪，人们大量使用添加氮、磷和钾的合成化肥（Tilman 等，2002）。在全球经济增长的支持下，化肥需求量持续增加（图6），从而导致了"越多越好"的过度施肥。据联合国粮农组织预测，到2018年全球化肥消费将达到2亿吨，其中有超过50%的化肥消费集中在中国、美国和印度（FAO，2015a）。然而，农业土壤施肥量的增加与作物产量的增加之间并没有线性正相关关系；

相反，增加化肥使用量可能导致养分利用效率低下，降低作物产量（Hossain 等，2005；Zhu 等，2005），并可能造成严重的环境问题（Good 和 Beatty, 2011；Vitousek 等，2009；Withers 等，2014）。

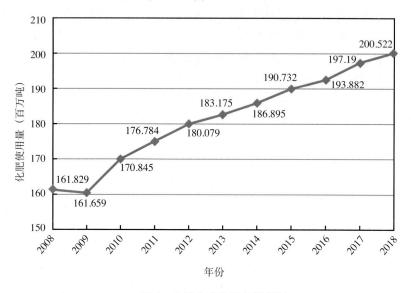

图6　全球合成化肥使用情况
资料来源：FAO，2015a。

当氮和磷以化肥的形式被过度地施用于农业土壤或牲畜集约化生产地区，它们就会成为污染物（Carpenter 等，1998；Torrent, Barberis 和 Gil-Sotres, 2007）。这些营养物质能够渗入地下水或通过径流输送到地表水体，造成富营养化或导致硝酸盐浓度过高等环境问题和人类健康问题（EC,1991；Frumin 和 Gildeeva, 2014；Pretty 等，2003；Yaron, Dror 和 Berkowitz, 2012）。在磷肥和氮肥中也发现了许多重金属，例如As、Cd、Cr、Hg、Pb和Zn（Brevik, 2013）。

虽然营养物质对作物生产至关重要，但过量施用就会对作物产量产生负面影响。氮增加了叶绿素的产量，造成花和根生长所需的能量被运输到叶片，促进叶片增殖生长，导致植物生长系统紊乱，更容易受到病原体的攻击。它还会影响作物的营养平衡（Hao 等，2003）。与此同时，氮素污染可以影响土壤的酸度和盐碱度（Han 等，2015），并影响到微生物群落的组成和活性，进而影响土壤有机质分解过程（Bragazza 等，2006；Luo 等，2017；Shen 等，2010；Zhou 和 Zhang, 2014）。

1.4.3　农药

农药的使用是为了减少因虫害、杂草和病源造成的作物损失，从而保证全球粮食供应（FAO 和 ITPS，2017）。农药包括但不限于杀虫剂、杀菌剂、除草剂、杀鼠剂、杀线虫剂和植物生长调节剂。

在不使用农药的情况下，作物损失估计会达到从谷物产量的32%到水果产量的78%不等（Cai, 2008）。农药不仅适用于农田，它们对人类健康的保护也非常重要，例如利用农药对虫霉病进行的卫生控制。它们还被用于使基础设施免受害虫和杂草的伤害，例如防止白蚁对木质建筑的腐蚀，或者保持路边和火车轨道的清洁从而有助于避免事故发生（Aktar, Sengupta 和 Chowdhury, 2009）。农药在全球的使用并不均匀，主要是由于其费用以及虫害因气候和地理区域的不同而不同。联合国粮农组织数据库提供了大量统计数据（FAOSTAT，2016），表明一些低收入和中等收入国家在过去十年中增加了农药的消费。例如，孟加拉国农药使用量增加了4倍，而卢旺达和埃塞俄比亚增加了6倍多。在苏丹，这个数字高达10倍。当农药的使用量超过了需求量，并采用了导致其容易扩散到环境中的做法，如使用不合适、未维护、未校准的设备喷洒或用飞机大面积区域喷洒，就会对居民和非靶向生物产生影响（Carvalho, 2017）。

农药可以是有机合成分子，也可以是无机合成分子。它们是根据其化学结构、作用方式、进入人体的方式以及靶向生物来分类的。它们对害虫的毒理学作用取决于它们的化学成分，而化学成分反过来又影响它们与土壤成分的相互作用（Singh，2012）。农药按其化学结构可分为12类，每一类的主要农药种类如下：

• 有机氯化物类：滴滴涕、甲氧滴滴涕、氯丹、三氯杀螨醇、六氯苯/六氯环己烷、艾氏剂、硫丹、七氯；

• 有机磷化物类：对硫磷、马拉硫磷、久效磷、毒死蜱、喹硫磷、甲拌磷、二嗪农、杀螟硫磷、甲胺磷、乐果、硫磷、异丙胺磷、磷胺磷、双硫磷、三唑磷；

• 氨甲酸酯类：涕灭威、氨基乙二酰、西维因、卡巴呋喃、丁硫克百威、灭多威、甲硫威、残杀威、抗蚜威；

• 拟除虫菊酯类：丙烯菊酯、溴氰菊酯、苄呋菊酯、氯氰菊酯、苄氯菊酯、氰戊菊酯、除虫菊酯；

• 新烟碱类：啶虫脒、吡虫啉、烯啶虫胺、噻虫嗪；

• 有机锡化物类：三苯基醋酸锡、三丁基氯化锡、三环己基氢氧化锡、三唑锡；

• 有机汞化合物类：氯化乙基汞、苯基溴化汞；

• 有机硫杀菌剂类：代森锌、代森锰、代森锰锌、福美锌；

• 苯并咪唑类：苯菌灵、多菌灵、甲基托布津；

• 氯苯氧基化合物：2,4-二氯苯氧乙酸、四氯二苯并-p-二噁英、敌草索、2,4,5-涕丁酸、2,4-二氯苯氧丁酸、二甲四氯、2-甲-4-氯苯氧基丙酸；

• 联吡啶：百草枯、敌草快；

• 其他类：二硝基甲酚、溴苯腈、西玛津、唑蚜威。

上面列出的一些农药也属于持久性有机污染物，下文将进一步讨论。

一些农药也与土壤重金属污染有关。最近，在政府间土壤技术小组关于植物保护剂对土壤功能和生态系统服务的影响报告中强调了含铜杀菌剂对蚯蚓和微生物量存在严重影响。而这些农药在有机葡萄栽培中被广泛用于葡萄真菌病的控制（FAO 和 ITPS，2017）。

由于农药在土壤中降解和保留机制非常复杂，导致其持久性、环境行为和流动性也千差万别（图7）。其在土壤中可经历吸附、解吸、挥发、化学和生物降解、植物吸收和淋溶等多个过程（Arias-Estévez 等，2008）。

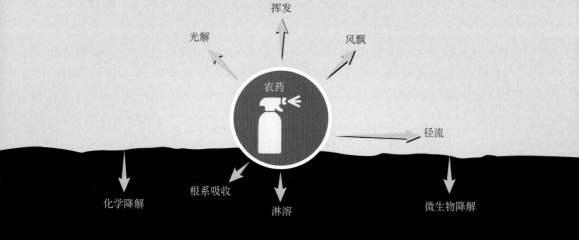

图7 杀虫剂在环境中的行为
资料来源：Singh, 2012。

1.4.4 多环芳烃

多环芳烃（Polycyclic Aromatic Hydrocarbons, PAHs）是一类持久性、半挥发性的有机污染物。

多环芳烃是由两个或多个未取代的苯环在共用一对碳原子时融合而成的一大类物理化学性质不同的分子。最常见的多环芳烃是蒽、荧蒽、萘、芘、苯并芘（Lerda, 2011）。多环芳烃极低的水溶性和固相传质速率限制了微生物对它们的利用率，进而阻碍了微生物对它们的自然衰减作用。多环芳烃由于其持久性和疏水性的特性使其在土壤中容易累积，并且可以在土壤中长时间存留。因此，大多数多环芳烃是持久性有机污染物的组成成分，其广泛存在于空气、水、土壤和沉积物中（Lin等，2013）。有两个或三个环的低分子量的多环芳烃具有挥发性，主要存在于大气中；而中等和高分子量的多环芳烃则根据温度变化以气态和固态平衡的形式存在（Srogi, 2007）。

煤炭、天然气、石油和垃圾的不完全燃烧；工业、农业和交通对有机材料的高温分解；天然有机质的成岩蚀变过程；长期废水灌溉；污水污泥再利用；农业生产中化肥的使用等。它们都会导致农业土壤中多环芳烃的浓度过高（Conte等，2001）（图8）。例如，在德国西部森林中，褐煤露天开采点已被鉴定为低分子量多环芳烃的主要来源（Aichner 等，2013），同时，Khalili等人认为二环和三环多环芳烃是多种排放源的主要产物，例如炼焦炉、柴油和汽油发动机以及木材燃烧（Khalili, Scheff 和 Holsen, 1995）。而交通排放和化石燃料燃烧被认为是城市地区多环芳烃的主要来源（Fabiaska 等，2016；Keyte, Harrison 和 Lammel, 2013）。

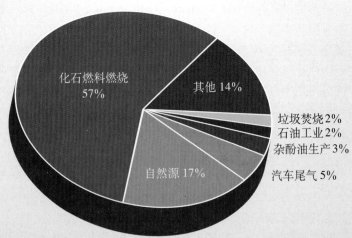

图8　多环芳烃全球排放源
资料来源：Kuppusamy 等，2017。

多环芳烃具有高毒性、致突变性、致癌性，并且广泛存在于环境中，因此受到人们的关注[欧洲食品安全局（EFSA），2008]。虽然多环芳烃种类非常多，但科学家和监管机构已经把重点放在16种已被确认为致癌物的化合物上（EC，2011；US EPA，1984）。然而，最近许多利益相关者支持实施更广泛的监管规定，其中不仅包括其他有毒的多环芳烃，还包括杂环芳香族化合物和烷基衍生物（Andersson 和 Achten，2015）。在对土壤中多环芳烃分布格局的全球分析中，Wilcke发现德国和捷克的多环芳烃污染比已经有相关研究的世界其他所有地区（如中国、俄罗斯、泰国、美国、巴西和加纳）都要严重（Wilcke，2007）。在最近的一项研究中，Loganathan和Lam发现，印度土壤中多环芳烃的浓度高于非洲、伊朗、巴西、俄罗斯、加拿大和澳大利亚土壤中的浓度。因此，多环芳烃是土壤中普遍存在的污染物，但其浓度随着与污染源的距离、土壤性质和气候条件的不同而变化。

进入土壤的多环芳烃可通过一系列物理化学和生物过程被稀释或降解，例如通过挥发或光氧化而进入到大气、不可逆地吸附到土壤有机质中、淋溶到地下水中、非生物性损失（受日和季节温度波动的影响），以及被植物吸收或微生物降解（Okere，2011；Šmídová 等，2012）。多环芳烃在未加工食品中的含量非常低，它们的浓度受其在水和有机溶剂中的相对溶解度控制。多环芳烃容易在植物和动物的脂质组织中累积，但不易在含水量较高的植物组织中累积。一般来说，多环芳烃从土壤转移到根茎类蔬菜中的量是非常有限的（Abdel-Shafy 和 Mansour，2016）。

1.4.5　持久性有机污染物

持久性有机污染物（Persistent Organic Pollutants，POPs）是在环境中可以持续存在，通过食物链会造成生物积累，对人类健康和环境会产生不利影响的一类化学物质（UNEP，2001）。持久性有机污染物种类有数千种，同时它们的来源丰富，广泛地被用于农业、疾病控制、制造业和许多工业等过程。持久性有机污染物包括氯化和溴化芳香烃，如多氯联苯，它们在各种工业过程中都有应用，例如在变压器和大型电容器中作为液压和热交换液，或者作为涂料和润滑剂的添加剂；还有有机氯杀虫剂，如滴滴涕及其代谢产物，这些农药在世界上一些地区仍然被用来控制携带疟疾的蚊子。一些人类活动无意产生的其他副产物，如由一些工业过程以及市政、医疗和家庭废物焚烧产生的二噁英（多氯代二苯并-对-二噁英和呋喃），也属于持久性有机污染物这一类（US EPA，2014b）。

持久性有机污染物主要是疏水性和亲脂性化合物，其与有机物和细胞质膜有很强的亲和性，所以可以被储存在脂肪组织中（Jones 和 de Voogt，1999）。

土壤污染：一个隐藏的现实

《斯德哥尔摩公约》是一项保护人类和环境免受持久性有机物污染的全球性协议，迄今为止已经列出了20多种持久性有机物（Stockholm Convention，2018）。持久性有机污染物通过在生物体脂肪内累积而进入食物链，并在从一个生物体转移到下一个生物体的生物富集过程中变得更集中（Vasseur 和 Cossu-Leguille，2006）。与此同时，持久性有机污染物也有很高的流动性：在天气温暖时，它们可以很容易地穿透水，进而从土壤挥发到大气中。当温度下降后，它们会在离释放点几英里远的地方沉降下来（Schmidt，2010）。通过流动造成持久性有机污染物污染的例子也有许多，如研究人员在北极的偏远区域发现了大量的持久性有机物[北极理事会北极监测和评估计划（AMAP），1997；Muir 和 de Wit，2010]。一般来说，分子氯化程度越高，其水溶性和挥发性就越低。多氯联苯不易被植物吸收，但容易在生物体内累积，主要存在于脂肪组织和母乳中（Passatore 等，2014）。

自从《寂静的春天》出版以来（Carson，2002），大量的研究都集中在持久性有机污染物对生物体和环境的影响方面（de Boer 和 Fiedler，2013；Cetin，2016；Muir 和 de Wit，2010；Mwakalapa 等，2018；Prestt, Jefferies 和 Moore，1970；Ratcliffe，1970；Vasseur 和 Cossu-Leguille，2006）。然而，在发展中国家，土壤中持久性有机污染物的存在情况仍然缺乏相关资料（Fiedler 等，2013）。考虑到亚洲发展中国家过去对城市废物管理不善和过量使用这类化学品的情况，持久性有机污染物在弃渣场造成的污染估计会很严重（Minh 等，2006）。

到20世纪70年代末，大多数政府已禁止多氯联苯的生产，但由于运输、储存和处置不当造成的意外泄漏和渗漏，其所导致的广泛环境污染仍然存在（图9）（Jones 和 de Voogt，1999；Passatore 等，2014）。尽管自《斯德哥尔摩公

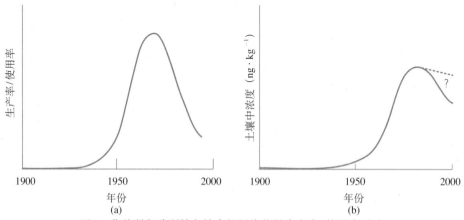

图9 北美洲和欧洲持久性有机污染物的生产率/使用率（a）
和持久性有机污染物在土壤中的浓度（b）
资料来源：Jones 和 de Voogt，1999。

约》通过以来，持久性有机污染物的使用和生产已经大大减少，但它们仍然存在于环境中，对粮食安全、健康和环境可以产生世世代代的影响（Odabasi 等，2016）。因此，为了防止未来农作物和动物受到污染，必须达成全球协议，加强努力和合作，清除土壤中的持久性有机污染物。

土壤是这些持久性污染物在环境中的主要载体。在土壤中，持久性有机污染物与土壤有机质形成稳定的化学键，以不可提取的形式存在。然而，土壤环境的一些变化会改变持久性有机污染物在土壤中的分配率，从而使它们易于被提取。例如，气温每升高1摄氏度，土壤中持久性有机污染物的挥发率就会增加8%（Komprda 等，2013）；低温有利于持久性有机污染物的沉积（Guzzella 等，2011）。另外，森林土壤由于有机碳含量高，也有利于长期累积持久性有机污染物（Kukučka 等，2009）。这些持久性有机污染物由于温度梯度所引起的纬向分布的结果（图10）被描述为"全球蒸馏效应"（Wania 和 MacKay，1996）。

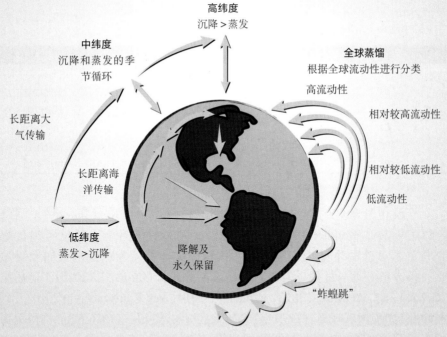

图10　POPs迁移过程
资料来源：Wania 和 MacKay，1996。

全氟和多氟烷基物质（PFAS）代表了一类人造化学物质，近年来因其在环境中含量丰富而备受关注。它们已经被证实存在于全世界范围内的土壤、水和人类血液中（Giesy 和 Kannan，2001；Kannan 等，2004；Rankin 等，2016）。

全氟和多氟烷基物质在过去的几十年里广泛应用于军事目的（如消防泡沫和金属电镀）(Lindstrom, Strynar 和 Libelo, 2011)，以及商业用途（如 Gore-Tex™ 和 Teflon™ 的生产）(Bossi, Dam 和 Rigét, 2015)。全氟和多氟烷基物质包括多种化合物，它们抗降解性强，并可以在食物链中积累，产生生物富集效应(Giesy 和 Kannan, 2001；US EPA, 2014c)。全氟和多氟烷基物质于2009年被列入关于持久性有机污染物的《斯德哥尔摩公约》，因为它们对人类健康存在潜在有害影响并且它们在生物体中具有强持久性，据估计它们在实验室大鼠身上的代谢时间为100天，而在人类身体上则可以超过5年(Wang 等, 2009)。

1.4.6 放射性核素

在环境中存在的放射性核素既可能是自然存在的，也可能是来自人为产生的物质（Mehra 等, 2010；Navas, Soto 和 Machín, 2002）。放射性核素的半衰期较长（表2），活性原子衰变期间发射的电离辐射是放射性核素的主要污染途径。

表2 土壤中主要放射性核素的特征

同位素	半衰期（年）	主要辐射	主要发生方式
^{14}C	5.7×10^3	β^-	天然和核反应堆
^{40}K	1.3×10^9	β^-	天然放射
^{90}Sr	28	β^-	核反应堆
^{134}Cs	2	β^-, γ	核反应堆
^{137}Cs	30	β^-, γ	核反应堆
^{239}Pu	2.4×10^4	α, X射线	核反应堆

资料来源：Zhu 和 Shaw, 2000。

土壤中最常见的自然源和人为源放射性核素有 ^{40}K、^{238}U、^{232}Th、^{90}Sr 和 ^{137}Cs (Wallova, Kandler 和 Wallner, 2012)。人为的核污染来源包括20世纪中期核武器试验的全球大气沉降物、核设施和非核工业的运营（如燃煤发电厂、核废料处理与处置以及放射性矿石的开采）(Ćujić 等, 2015)、矿质肥料 (Schnug 和 Lottermoser, 2013；Ulrich 等, 2014；Van Kauwenbergh, 2010)，以及核事故 [如美国三哩岛（1979年）、乌克兰切尔诺贝利（1986年）、巴西戈亚尼亚（1987年）、日本东海村（1999）和福岛（2011）]。

土壤中的放射性核素被植物吸收，从而可以在食物链中进一步地再分配(Zhu 和 Shaw, 2000)。例如，福岛核事故后，为了确保食品安全，对农产品进行了严格的监控（Nihei, 2013）。监测结果显示，蔬菜产品中的放射性核素含量迅速衰减，同时也发现，土壤中的放射性核素在最初污染后很长时间内仍然具有生物可用性（Absalom 等, 1999；Falciglia 等, 2014；Yablokov, Nesterenko

和 Nesterenko，2009）。虽然强烈建议在发生重大放射性事故后清除表土，但是大范围内清除表土不太可能实现，因为会产生大量的放射性废物。由于这个原因，农业区往往被废置多年。为了减少放射性核素在食物链中的转移，并促进可能受影响的土壤恢复其农业用途，必须采取相应的农业对策（Vandenhove 和 Turcanu，2011）。放射性核素向动物源性食品的转移已经存在一些相关研究分析（Howard 等，2009；Štrok 和 Smodiš，2012），但是，其中的机制还没有完全清楚或不是很好理解。

1.4.7 新型污染物

新型污染物（Emerging Pollutants，EPs）是指最近在环境中出现的大量合成的或自然存在的且通常没有被监测的化学物质（Geissen 等，2015）。它们具有可以进入环境并造成已知或疑似有害生态或人类健康的潜能。新型污染物很可能成为新出现的、令人关注的污染物，因为最新的事实或信息表明，它们会对环境和人类健康构成威胁（Sauvé 和 Desrosiers，2014）。新出现的污染物包括化学类物质（如药物、内分泌干扰物、激素和毒素）和生物污染物（如土壤中的微生物，包括细菌和病毒）。

自20世纪70年代以来，全球化学品的人为生产量快速增长。2016年，欧盟化学工业生产了3.19亿吨危险和非危险化学品。其中，有1.17亿吨被认为是对环境有害的[欧盟统计局（EUROSTAT），2018]。预计到2030年，全球化学品的生产量将以每年约3.4%的速度增长（图11），且未来非经济合作与发展组织国家对其贡献将更大[经济合作与发展组织（OECD），2008]。过去十年来，危险化学品的生产和使用有所减少；然而，由于仍然存在不确定性，而且缺乏许多发展中国家的相关信息，因此很难得出环境和人类健康风险已成功降低的结论。

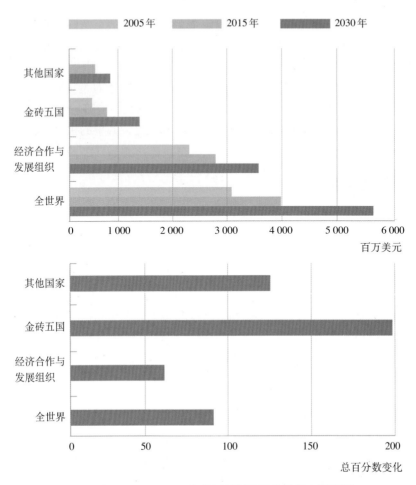

图11　2005—2030年按区域划分的化学品产量预测
资料来源：经济合作与发展组织，2008。

作为例子，药物的吸附行为等性质在不同的土壤类型中会有很大的差异，因为它们以电离和非电离形式出现会直接影响它们与土壤中不同化合物间的相互作用（ter Laak 等，2006）。在环境中出现大量潜在有毒新型污染物的情况下，更好地了解它们的出现规律、去向和生态影响更为必要（Petrie, Barden 和 Kasprzyk-Hordern, 2015）。

由于新型污染物的来源多种多样，其性质、理化性质也各不相同。其中包括挥发性、极性、吸附性、持久性及其与环境的相互作用。现将影响农业土壤的主要新型有机污染物分类如下。

医药和个人护理产品（PPCPs）是一类新型化学污染物，它们已经被广泛使用了几十年。有超过4 000种医药和化学产品，包括医用药品、诊断用药、化妆品、香料、营养品以及用于许多家庭清洁用品的添加剂。许多PPCPs都具有生物活性（Boxall 等，2012），且被设计用来与激素或活体组织相互作用；因此，当PPCPs释放到环境中时，了解它们的去向、影响和潜在风险是很重要的。

自20世纪90年代末以来，世界范围内在环境基质中检测出PPCPs的报道越来越多（Buser, Poiger 和 Müller, 1999；Hamscher 等，2004；Heberer, 2002；Jones, Voulvoulis 和 Lester, 2001）。PPCPs会进入城市废水池进行处理，但传统的处理技术无法有效地消除它们（Miège 等，2009），从而最后存留在污泥中。市政有机固体废物中已经发现含有大量的PPCPs，而它们作为肥料在土地上施用可能会将这些化合物引入到环境中，对有益微生物造成危害并影响营养物质循环。虽然PPCPs在土壤中的持久性是显著的，但关于生物改良后的土壤中PPCPs浓度的相关研究资料非常有限（Wu, Spongberg 和 Witter, 2009；Wu 等，2010）。与PPCPs相关的另一个问题是抗菌剂的存在，及其在环境中促进细菌耐药性的潜力（Walsh 等，2003）。

长期接触PPCPs可能产生的影响仍然相对未知，但最终可能导致土壤和水生生物的慢性中毒（Chalew 和 Halden, 2009）。PPCPs已经被证实与抗生素耐药细菌的发展、雄性鱼的雌性化和水生生物的基因毒性有关（Daughton 和 Ternes, 1999）。当前，仍然需要进行大量的研究，以期能对PPCPs进行可靠的风险评估，并明确细菌抗药性和耐药性的产生机理（Walters, McClellan 和 Halden, 2010；Sun 等，2018）。联合国大会已经认识到解决抗生素耐药性并减少土壤中抗生素残留问题的必要性[联合国（UN），2016）]。

环境中抗生素残留的出现及其影响正在成为一个新型关注点。人类每天都要服用抗生素、杀菌剂和其他药物，为了促进家畜的生长并减少或预防疾病，这些药物也广泛用于家畜。众所周知，在用药后，药物会被吸收并发生代谢反应（如羟基化、裂解或葡萄糖醛酸化）产生代谢物，这些代谢物可能比原来的化合物更有害，或者可能会转化为原来的活性化合物（Díaz-Cruz 和 Barceló, 2005）。大部分药物没有被吸收或代谢，而是通过粪便或尿液排出体外。因此，药物不断被排入城市废水和粪肥中。当粪肥和污水污泥作为肥料施用于农田，或处理过的废水用于农业灌溉时，作物会暴露在抗生素中，这些抗生素可能会在土壤中存留数天到数百天。有文献证明，某些抗菌药物，特别是抗生素、阿莫西林和氟喹诺酮类药物，可以被作物吸收（Azanu 等，2016）；而其他的PPCPs，如咪康唑（杀菌剂）和氟西汀（抗抑郁药），尽管它们在土壤中持续存在，但它们在植物中并没有被检测到（Gottschall 等，2012）。

增塑剂是用于增加柔韧性或可塑性的一种添加剂，如丙二酚（BPA）或邻苯二甲酸盐，同时它们被认为是一种内分泌干扰物（Ghisari 和 Bonefeld-Jorgensen, 2009）。增塑剂已经被禁止或受到严格的管制[美国联邦法规

(CFR), 2017; EC, 2006; 工业化学品申报和评估计划 (NICNAS), 1989; 台湾环境保护局 (TwEPA), 2014]。邻苯二甲酸酯被 (PAEs) 广泛用作增塑剂，并出现在许多产品中，如润滑油、汽车部件、油漆、胶水、驱虫剂、摄影胶片、香水和食品包装。在靠近城市或城郊地区的许多农业土壤中都能检测到PAEs和BPA的存在，它们来自污水污泥的应用、塑料薄膜的农业使用、城市废水的农业灌溉或大气沉降（Tran 等，2015；Zeng 等，2008）。邻苯二甲酸盐和丙二酚通过与内源性激素的特定受体结合或破坏它们的合成和代谢途径，与内源性激素产生竞争（Craig、Wang 和 Flaws，2011）。邻苯二甲酸盐和丙二酚都已经在食品和人体中被检测到，而且已经在国际法规中被列为有毒物质[澳大利亚政府 (Australian Government), 2018；EC, 2006；联合国欧洲经济委员会 (UNECE), 2011；US EPA, 2012；Yen、Lin-Tan 和 Lin, 2011]。

另外两大类新型污染物是人造纳米颗粒（Manmade Nanoparticles, MNPs）及其副产物。在过去的几十年里，含有或需要MNPs的产品数量急剧增加，它们出现在数千种产品中，包括作为添加剂出现在油漆、化妆品、纸张、塑料和食品中（Fiorino, 2010）。它们还被用于纺织品中，用来生产可以自洁、防水和防污、抗微生物、抗紫外线且耐磨的衣服。人造纳米颗粒也被用于土壤修复，来减少有机和无机污染物的影响，同时，它们也通过其他多种途径被无意地释放到土壤中（Pan 和 Xing, 2012）。

MNPs的特性以及与土壤基质、生物排泄物和微生物的相互作用（图12）目前仍然尚不清楚（Nel 等，2006）。主要是由于缺乏有关其性质的可用信息，如溶解度、物理结构、形状和表面电荷。人造纳米颗粒在进入环境前后的转变，如腐殖酸的表面改性、与阳离子的相互作用和溶解，可能影响它们在环境中的去向（Liang 等，2013）。因此，新型污染物风险管理仍然处于零散和停滞不前的状态（Geissen 等，2015）。当水处理（饮用水或废水）的试剂与基质组分反应产生新物质，或当目标污染物的反应不完全，就会产生一些可能有残留毒性的副产物，这时就应该对副产物进行处理（Handy 等，2008）。

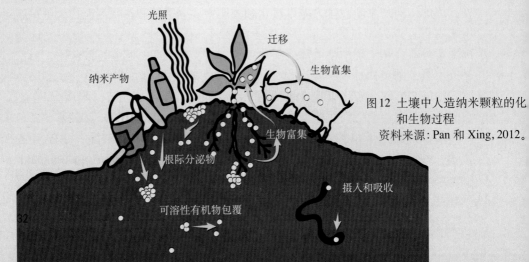

图12 土壤中人造纳米颗粒的化和生物过程
资料来源：Pan 和 Xing, 2012。

人造纳米颗粒主要通过四种机制对生物体产生毒性作用：生成活性氧化物质，引起氧化应激反应；改变膜细胞的渗透特性，干扰生理活动；改变电子传递过程；修复蛋白质构造，干扰生物信号的传递和基因形成（Pan 和 Xing, 2012）。

出于供应链安全和生物地球化学磷循环中断的担忧，人们正在寻求新的、更可持续的、适销对路的产品。解决这个问题的新途径是促进有机和废物来源肥料的使用，如磷酸铵镁（废水中沉淀的铵镁，是一种缓释肥料）、生物炭或灰。这些新的肥料产品旨在循环利用原本会流失的营养物质。然而，目前尚未建立起将这些产品安全地应用于土地和作物的质量标准。它们可能含有重金属和类似 PPCPs（激素等）的残留物，对环境有不利影响。它们的产量和销量预计将在未来 10 到 20 年出现全球性增长。当前欧盟正在通过采用循环经济的新计划来应对这一新的挑战（EC, 2015），并计划在新的欧盟肥料条例框架内对污染物的含量进行管制（Vollaro, Galioto 和 Viaggi, 2017）。

1.4.8 病原微生物

由于巨大的生物多样性和微生物量，每平方米土地有超过1万种细菌，每公顷面积有超过数吨的细菌生物量（欧盟委员会联合研究中心和全球土壤生物多样性倡议，2016），所以在土壤内部存在着对资源的巨大竞争。有些生物体可以通过排泄化合物杀死或干扰其他遇到这些化合物的微生物的生长，从而形成化学防御。在这些生物体中，绝大多数不会对人类健康构成威胁，而是通过土壤内部生物体之间或与土壤本身之间的多种复杂相互作用，提供大量的生态系统服务。然而，其中一些生物体可能会因为引起土壤传播疾病而对人类有害。它们要么作为机会性病原体，侵染易感个体（如免疫系统受损者），要么作为专性病原体，必须感染人类，以完成其生命周期（Van der Putten 等，2011）。在感染接触污染土壤的人类之前，其中一些生物体可能在土壤中存活很长一段时间，而另一些生物体则需要在离开之前的寄主后立即感染人类。

Van der Putten 等将土壤传播的人类疾病定义为"由病原体或寄生虫引起的人类疾病，这些病原体或寄生虫即使在没有其他感染性个体的情况下，也可通过土壤进行传播"。他们整理出了一份完整的病原体列表，并区分了真正的土壤来源生物体（土壤病原微生物），以及由于具有抗性结构从而能在土壤中存活很长时间的病原体（土壤传播病原体）（表3）。

表3 土传病害及其病原体

土壤优先寄生生物	土壤传播的病原体
放线菌瘤：衣氏放线菌	小儿麻痹病毒
炭疽：炭疽杆菌	汉坦病毒
肉毒中毒：肉毒梭菌	寇热病：伯纳特氏立克次氏体

(续)

土壤优先寄生生物	土壤传播的病原体
弯曲杆菌病：如空肠弯曲菌	莱姆病：包柔氏螺旋体
钩端螺旋体病：钩端螺旋体	蛔虫病：蛔虫
李斯特氏菌病：李斯特菌	钩虫病：十二指肠钩虫
破伤风：破伤风梭菌	蛲虫病：寸白虫
土拉菌病：土拉热杆菌	类圆线虫病：如粪类圆线虫
气性坏疽：产气荚膜梭菌	鞭虫病（鞭虫）：毛首鞭虫
肠道耶氏症：结肠炎耶尔森杆菌	包虫病：棘球绦虫
曲霉病：曲霉属真菌	旋毛虫病：旋毛虫
芽生菌病：皮炎芽生菌	阿米巴病：痢疾变形虫
球孢子菌病：粗球孢子菌	小袋虫病肠袋虫属杆菌
组织胞浆病：荚膜组织胞浆菌	隐孢子虫病：隐孢子虫
分支孢菌病：申克氏孢子丝菌	圆孢球虫病：圆孢子虫
毛霉菌病：根霉	梨形鞭毛虫病：鞭毛虫
足分支菌病：诺卡氏菌属	等孢球虫病：贝氏等孢子球虫
类圆线虫病：粪类圆线虫	弓形体病：弓形虫
	志贺氏菌病：志贺氏杆菌，铜绿假单胞菌，大肠杆菌
	沙门氏菌病：肠道沙门氏菌

资料来源：Van der Putten等，2011。

有些病原体可能来自动物排泄物，而土壤则是病原体通过皮肤接触或接触受污染的水和食物而进行传播的主要途径（Ercumen等，2017）。蠕虫是一种寄生虫，存在于人类粪便中，污染的主要是卫生条件差的地区的土壤。据世界卫生组织估计，全世界有20亿人感染了通过土壤传播的寄生虫（WHO，2001b）。使用未经处理的生活废水灌溉或者使用处理不当的粪肥改良土壤这些不安全的农业操作，在发展中国家的小农户非常普遍，同时也存在于部分发达国家（Allende和Monaghan，2015；Ongeng等，2011；du Plessis，Duvenage和Korsten，2015；Scallan等，2011）。由于健康饮食的重要性，不经过加工或低程度加工的水果和蔬菜（如生菜、菠菜和胡萝卜）数量正在显著增加。然而，当使用不适当的方式生产时，它们可能会成为肠道病原体的来源，这一点已被越来越多的与食用新鲜产品相关的人类感染事件报道所证明（Beuchat，2002；Ingham等，2004）。接触受污染的土壤已被确定为食品污染的一个潜在来源（Khandaghi，Razavilar和Barzgari，2010）。

1.4.9 抗生素抗性细菌和基因

细菌的遗传适应性很强，其会在抗生素的反复作用下发生突变，这种突变会导致细菌对抗生素产生耐药性。因此，抗生素的增加与广泛使用正在使环境中产生出更多的具有抗生素耐药性的细菌（Rensing 和 Pepper, 2006）。微生物产生抗药性的另一种方式是外来抗生素耐药基因的转移，或者来自农业活动引入的细菌（例如畜牧业、人类污水处理、不当堆肥肥料），又或者来自被驯化的和野生的动物排泄到土壤微生物群落的粪便（FAO, 2016）。抗生素耐药性是当前社会面临的一个重大问题：目前，在欧洲和美国由于抗生素耐药性感染每年夺去大约5万人的生命。据预测，到2050年，如果这个问题得不到解决，抗生素耐药性导致的死亡人数将超过癌症，在全球范围内造成的损失，将超过当前全球经济规模（O'Neill, 2014）。近年来世界范围内强耐药致病菌在微生物圈的富集和传播，很大程度上是由人类活动导致的，包括抗生素在人类医学、兽医学和农业生产中的广泛使用和滥用（Witte, 1998）。

抗生素常常用于牲畜的促生长、疾病的预防和治疗方面（Joy 等, 2013）。据估计，2010年全球畜牧业抗菌药物使用量为63 151吨（FAO, 2016）。然而，使用的抗生素有很大一部分没有被动物吸收（Sarmah, Meyer 和 Boxall, 2006）。一旦在土地上施用动物粪肥，粪肥的抗生素将在土壤颗粒对化合物吸附特性的影响下，通过其在土壤中的迁移转化和它们随后的径流输送形成影响（Sassman 和 Lee, 2005）。

1.5 污染物与土壤成分的相互作用

土壤提供的主要生态服务之一是过滤、缓冲及转化无机和有机污染物。这一基本功能为地下水和食品安全生产提供了保障（Blum, 2005）。当污染物进入土壤后，土壤会经过物理、物理化学、微生物和生物化学过程，来保留、减少或降解这些污染物。

影响污染物行为的重要土壤特征包括土壤矿物和土壤黏土含量（土壤质地）、土壤有机质含量、土壤酸碱度、土壤湿度以及温度。污染物本身的性质也非常重要，包括污染物的大小、形状、分子结构、溶解性、电荷分布和分子的酸碱性（Gevao, Semple 和 Jones, 2000）。

1.5.1 污染物的吸附

吸附是流体分子与固体相互作用并在固体上停留一段时间的过程（Navarro, Vela 和 Navarro, 2007）。吸附本质上可以是化学的（如离子和氢的结合），也可以是纯物理的（如范德华力）。

带正电荷的离子或带电分子（阳离子型）参与带电表面的阳离子交换。土壤有机质和黏土矿物是土壤中阳离子交换位点的来源，不同土壤有机质组分和不同型的黏土矿物的阳离子交换能力差异非常大。带负电荷（阴离子型）的离子或分子一般在土壤中结合较弱，主要通过氢键和配位体交换与土壤有机质反应（Gevao, Semple 和 Jones, 2000）。某些离子或分子的氧化态会随着土壤溶液的pH变化而转变为阳离子、中性或阴离子。因此，它们在土壤中的吸附依赖于土壤pH。此外，一些分子会得到或失去质子，因此表现出酸性或碱性行为，这些分子的吸附也依赖于土壤pH。最后，由于土壤湿度条件的变化，氧气的存在与否会发生变化（即土壤的氧化还原状态），也会引起一些离子和化合物氧化状态的改变，这也是控制这些污染物流动性的重要手段。

氢键是具有极性的非离子型农药（即分子中电荷的不对称分布）的重要吸附方式。许多农药都是非离子型和非极性的，并通过物理的范德华力与土壤有机质反应。范德华力对于能与表面紧密接触或能黏滞表面的离子作用是最强的。因此，农药的大小和结构是控制农药吸附的重要因素（Gevao, Semple 和 Jones, 2000）。此外，非极性分子不会被极性水分子吸引，通常不溶于水（即疏水），它们会与土壤有机质发生大量而复杂的反应，导致它们长期被固存。例如，滴滴涕具有高度不溶性，并且与土壤有机质有很强的亲和力，这解释了滴滴涕可以在环境中持久存在的部分原因（Mansouri 等, 2017）。

当前，在污染物吸附和土壤性质方面已经得到几条公认的结论。首先，土壤中有机质含量（特别是高活性腐殖质含量）是吸附的主要控制因素。土壤有机质既为离子反应提供了带电位点，又提供了可增强物理吸附过程的高度复杂结构。溶解的有机物也可与纳米颗粒相互作用，改变其表面性质和聚集状态，从而增加纳米颗粒的流动性和生物利用率（Pan 和 Xing, 2012；Wang 等，2011）。其次，黏土含量和黏土矿物的性质对吸附有很强的控制作用。全球研究表明，细粒土壤比粗粒土壤表现出更大的离子吸附倾向。这是由于细粒土壤含有表面积大的土壤颗粒，如黏土矿物、铁和锰的氢氧化物以及腐殖酸等

(Bradl，2004）。在比较温带土壤和热带土壤时，其所含的黏土矿物显得非常重要。热带土壤通常是高度风化的，具有低活性的黏粒，其电荷高度依赖于pH（Lewis 等，2016）。最后，根据前两种概括，有机质含量低的沙质土壤由于缺乏吸附位点而具有污染物淋滤的特殊风险。

1.5.2 污染物的生物有效性、流动性和降解

生物有效性是指对生物体暴露于土壤化学物质有重要影响的物理、化学和生物过程的相互作用（土壤和沉积物中污染物生物有效性委员会，2002）。

1.5.2.1 金属

土壤吸附能力对重金属和类金属的生物有效性有重要影响。土壤生物和植物只对离子形态的金属发生生物吸附。许多金属以简单的阳离子形式出现（表4），但像As和Cr等金属则会形成更复杂的含氧阴离子。金属可以吸附在土壤中的极细的有机物颗粒（腐殖质）、黏土矿物、铁锰氧化物和一些低溶性盐类如碳酸钙的表面（Morgan，2013）。在黏土矿物和放射性核素之间也观察到类似的关系（van der Graaf 等，2007）。金属还通过与有机分子的相互作用形成复杂的化合物；铜对于形成这种化合物具有特殊的亲和力（Morgan，2013）。

许多金属的吸附过程与pH有关。非酸性土壤的吸附力最强，而酸性条件有利于金属解吸并释放回溶液中。由水饱和引起的厌氧条件也会导致某些金属的解吸。

通过添加有机和无机改性剂可增加结合位点的数量和改变土壤的pH，对降低土壤中重金属的生物利用率非常有效（Puschenreiter 等，2005）。这些改良剂包括堆肥、生物固体（污水污泥）、粪肥和工业活动的副产品。这些措施可以对环境产生许多积极影响，同时有助于减少废物。施用石灰会增加土壤的pH，减少作物对金属的吸收（Knox 等，2001）。

表4　土壤金属污染物在土壤中的主要存在形态

元素	符号	土壤中的主要形态
砷	As	AsO_3^{2-}、AsO_4^{3-}
镉	Cd	Cd^{2+}
铬	Cr	Cr^{3+}、CrO_4^{2-}
铜	Cu	Cu^{2+}
汞	Hg	Hg^{2+}、$(CH_3)_2Hg$
镍	Ni	Ni^{2+}
铅	Pb	Pb^{2+}
锌	Zn	Zn^{2+}

资料来源：Logan，2000。

1.5.2.2 放射性核素

黏粒含量、pH和土壤有机质对土壤中^{137}Cs以及其他放射性核素的吸附也起着重要作用（Absalom，Young 和 Crout，1995；Rigol，Vidal 和 Rauret，2002）。土壤中的微生物通过催化化学转化过程作用于放射性核素的地球化学循环（Turick，Knox 和 Kuhne，2013）。

Kd分布（或分配）系数用来描述放射性核素被吸收的倾向，低Kd值表示对放射性核素的吸收较低。

铯的化学反应活性不强，在土壤中的行为与钾相似。它在低黏土含量的土壤和高岭石含量的土壤中具有较高的生物有效性（Kd=240～290），但是在伊利石黏土矿物表面可被强烈地束缚（Kd=6 300～8 300）。

碘具有多种氧化还原状态，因此表现出复杂的行为。碘的-1、+5价态和分子态（I_2）在环境中最为常见。碘在矿质土的Kd值在0.04至81之间（Turick，Knox 和 Kuhne，2013），在矿质土中的整体吸附量较低。碘的吸附一部分由土壤有机质控制，一部分由铁和铝的氧化物控制，在酸性越强的环境下，其吸附量越多。

从生物学角度来看，铀的+4和+6价状态是最重要的。+4价态的铀大部分是不溶的，而且大部分是不具流动性的，而+6价态的铀在环境中是可溶的且具有可流动性。铀的吸附对pH有很强的依赖性。例如铀与矿物磷灰石反应时，pH为4时Kd值为668，pH为7时Kd值为24 660（Turick，Knox 和 Kuhne，2013）。

钚具有复杂的地球化学性质，可以以+3、+4、+5和+6价氧化态存在。它通常很容易与土壤基质表面结合，从而被固定。它的吸附过程也依赖于pH，且在pH为6时出现最大吸附（Turick，Knox 和 Kuhne，2013）。

1.5.2.3 农药

不同农药的化学成分和结构以及它们与土壤成分的相互作用，存在着很大的差异（Gevao，Semple 和 Jones，2000）。分布系数（Kd）为土壤吸附的农药浓度除以溶液中的浓度。Kd值越高，说明农药的吸附力越强。在许多土壤中，有机质的含量在很大程度上控制着土壤的吸附能力，吸附系数（Koc）的计算方法是用Kd值除以土壤中有机碳的含量。Koc值越高，化学物质的吸附力越强，在环境中的流动性越小。一般来说，像草甘膦、硫丹，尤其是滴滴涕等，具有较高的Koc值，其流动性相对较小（表5），但这并不是绝对的，还会受到其他因素的影响。例如，Dores等人观察到，尽管硫丹具有较高的吸附系数，但他们检测的巴西土壤中还是发现了硫丹有淋溶现象（Dores 等，2016）。他们把这归因于土壤的优先流。

表5 文献中报道的不同农药的吸附系数（Koc）值

农药种类	Koc mL·g^{-1}	农药种类	Koc mL·g^{-1}
2,4-D	20～32	α-硫丹②	8 725～31 992
草脱净	163～172	β-硫丹②	8 186～31 992
克百威	29.4	草甘膦①	10 891～14 863
草克乐	98～107	林丹	1 081～1 340
毒死蜱②	1 671～2 896	马拉松	20
百草敌	2.2	百草枯	20 000
滴滴涕	243 000		

注：①来自Farenhorst等，2008；②来自Dores等，2016；其他来自Wauchope等，2002。

微生物的多样性和活性，特别是细菌和真菌，也对污染物的生物有效性有重要影响。微生物能够降解和转化某些污染物，释放出副产物，并影响毒性和流动性（Burgess，2013）。虽然大多数农药都是自然界中原本不存在的新结构，但有些仍可以被微生物代谢掉（Topp，2003）。一旦能够降解农药的微生物被识别出来，就可以在生物修复过程中通过接种这种微生物来净化被污染的土壤。

有氧或无氧条件也被证明对污染物的持久性和生物有效性有显著影响。例如，Ying等观察到三氯二苯脲和三氯生这两种普遍存在于个人护理产品中的抗菌剂，其生物降解过程只能发生在有氧条件下，而在厌氧条件下其持久性要长得多（Ying，Yu和Kookana，2007）。脱氯细菌在生态系统中可以降解含氯化合物，而此过程通常需要一定的时间使其适应目标化学物质（Brahushi等，2004）。在荷兰的两个相邻污染土壤中观察到的结果显示，这种适应在污染物浓度高时似乎发生得更快（Middeldorp等，2005）。

1.5.2.4 持久性有机污染物

持久性有机污染物是由于其具有强吸附性、疏水性或抗微生物降解结构等因素而在土壤中很难被降解的一类物质。多氯联苯具有疏水性、非极性和惰性，二噁英和呋喃也是如此。对于多环芳烃，其疏水性和化学反应活性随分子量的增加而降低，因此，他们在环境中的周转过程很难统一描述（Burgess，2013）。

1.5.2.5 氮磷肥

关于氮的不同形式及其生物利用率和流动性的特性已经得到了很好地认定（Cameron，Di和Moir，2013）。土壤中的氮主要以四种形式存在：①土壤有机质；②土壤生物体和微生物；③结合在土壤颗粒上的铵离子（NH_4^+）；④土壤溶液中的矿物氮形态，包括铵盐（NH_4^+）、硝酸盐（NO_3^-）和少量亚硝酸盐（NO_2^-）（Cameron，Di和Moir，2013）。在通气良好的土壤中，可供植物吸收的氮主要是硝酸盐形态，而在湿地或酸性土壤中主要是铵盐形态（Krapp，2015）。

硝酸盐是一种阴离子化合物，其吸附性较差，土壤中硝酸盐的淋溶会引起肥力的丧失，而且会对环境和人类健康构成威胁。饮用水中的硝酸盐被证明与婴儿的高铁血红蛋白血症以及癌症和心脏病有关（Cameron, Di 和 Moir, 2013）。与此同时，氮素也可以以气体的形式从土壤中流失。氮肥、动物尿液和粪便在农田中的使用会引发氨挥发，具有强温室气体效应的 N_2O 在氮循环的多个过程中产生，其中最大的排放过程来自饱和土壤的反硝化作用。

土壤中的磷主要以正磷酸盐（$H_2PO_4^{2-}$、$H_2PO_3^-$）和有机磷形态为主；它也可以以吸附态出现，如与铝、铁氧化物、其他矿物质和有机物形成表面复合物。磷可以被固相强烈地保留，并通过侵蚀的固体颗粒、粪便及人类废物的形式被传输（Yuan 等, 2018）。由侵蚀和径流造成的农田磷损失是水生系统中磷的主要贡献来源：据 Yuan 估计，从农田到淡水的磷损失为 10.4±5.7 百万吨/年，由侵蚀造成的自然磷损失为 7.6±3.3 百万吨/年。Civan 等人在研究英国河流中的磷来源时发现，点源磷污染通常更多地以正磷酸盐和生物可用磷形态存在，而来自农场径流、猪粪尿和农田侵蚀等扩散源中的磷通常会被吸附到颗粒物中（Civan 等, 2018）。总的来说，他们发现自 1985 年达到峰值以来，英国河流的平均总活性磷浓度下降了 60%，他们认为这是污水处理厂采取措施的成果。

2 土壤污染对食物链和生态系统服务的影响

预计到2050年世界人口将超过90亿，人口的增长对高质量食物和水的供应提出了更高需求（Godfray 等，2010；McBratney, Field 和 Koch, 2014）。据Dubois的相关研究指出，与2009年的产量水平相比，到2050年全球粮食产量将增长70%，发展中国家将增长100%。联合国粮农组织的最新预测表明，全球粮食产量将在2005年7月至2050年间增长60%。这是在最新数据和信息基础上，对2009年同期预测（70%）的下调。（World Agriculture Towards 2030/2050: The 2012 revision ESA E Working Paper No. 12-03 http://www.fao.org/economic/esa/esag/en/，走向2030/2050年的世界农业；2012版。ESAE工作文件12-03号）。食物的数量和营养质量支撑着人类的健康，而95%的食物生产依赖于土壤（Oliver 和 Gregory, 2015；FAO, 2015）。只有健康的土壤才能提供所需的生态系统服务，并确保供应更多的食品和纤维制品。生态系统所提供的服务受到了非常多的关注，其可以被定义为"自然过程及其成分直接或间接地提供满足人类所需求的商品和服务的能力"（Groot, 1992）。粮食安全被定义为"粮食的可获取性、流通性、可利用性和供应的稳定性"。土壤污染会通过两方面影响粮食安全，一方面会降低作物产量；另一方面会导致生产的作物不安全而无法食用（FAO 和 ITPS，2015）。

2.1 土壤污染、植物吸收和食物链污染

土壤污染物通过植物转移到食物链中的途径如图13所示。

图13　植物对土壤污染物的主要吸收途径
资料来源：改编自Collins, Fryer和Grosso, 2006。

如果一种污染物在低浓度下对植物有很高的毒性，但是不易转移到芽、果实或块茎上对动物和人类造成危害，那么它就不太可能进入食物链并成为一种危害。Chaney在大约40年前，把这个概念称为金属和类金属的"土壤—植物屏障"（Soil-Plant Barrier；Chaney, 1980）。根据对人类健康的危害，Chaney定义了当污水污泥施用于土壤时会进入食物链的四类金属（表6）。

表6　金属／类金属通过植物吸附造成潜在食物链风险情况分类

第1类	第2类	第3类	第4类
银（Ag）	汞（Hg）	硼（B）	砷（As）
铬（Cr）	铅（Pb）	铜（Cu）	镉（Cd）
锡（Sn）		锰（Mn）	钴（Co）
钛（Ti）		钼（Mo）	钼（Mo）
钇（Y）		镍（Ni）	硒（Se）
锆（Zr）		锌（Zn）	铊（Tl）

资料来源：改编自Chaney, 1980。

第一类包含的元素为不能被植物吸收，对食物链污染风险较低的元素。因为它们在土壤中的溶解度低，所以被植物吸收和转移的可能性很小。在食物中这些元素浓度升高通常表明是土壤或粉尘的直接累积污染。第二类包含的元素为可以被土壤表面强烈吸收，且可能会被植物根系吸收，但是却不易被转移到可食用组织中的元素，因此对人类健康造成的风险极小。然而，如果被污染的土壤被直接摄入，这些元素可能会对食草动物（或人类）构成一定风险。第三类包含的元素是容易被植物吸收的元素，但其只对植物造成毒性影响，其浓度对人类健康无影响。从概念上讲，"土壤—植物屏障"保护食物链免受这些元素的污染。第四类由对食品链污染风险最高的元素组成，它们在一般植物组织浓度下对植物无毒性但却对人类或动物健康构成风险。Chaney最初将砷归于第二类，但过去20年的研究表明，淹水土壤中的低氧化还原条件使得水稻种植农田面临着砷通过食物链转移的风险。这样的条件提高了砷的溶解度和被水稻吸收的能力，因此现在砷被列为高危的第四类元素。砷和镉对土壤的污染可能是现存的对全球食物链最普遍的风险（Grant等，1999；McLaughlin, Parker和Clarke, 1999），在东南亚大片地区的土壤都受到砷和镉的污染（Meharg, 2004；Hu, Cheng和Tao, 2016）。

在中国的一些地区，被重金属污染的土壤仍然会被用来种植作物。而在这些土壤中生长的作物通常会被重金属污染。根据"中国对话"，估计每年有1 200万吨被污染的粮食必须被处理掉，给中国农民造成200亿人民币的损失，约合25.7亿美元（Lynn, 2017）。

土壤中砷、镉、铅和汞等过量重金属也会损害植物代谢，降低作物产量，从而最终对可耕地造成压力。当这些重金属进入食物链后，它们还对粮食安全、水资源、农村生计和人类健康构成危害。金属在地上组织中的吸收和转运受植物遗传和生理差异的制约（Chen, Li和Shen, 2004），还受土壤中金属浓度及暴露时间影响（Rizwan等，2017；Tőzsér, Magura和Simon, 2017）。一旦重金属元素进入植物组织，它们可能会干扰植物的多个代谢过程，降低植物生长，产生毒性，并最终导致植物死亡。已报道的主要影响有降低发芽率、氧化损伤、降低根和芽伸长率以及改变糖和蛋白质的代谢（Ahmad和Ashraf, 2011）。例如，高浓度的铅可以加速活性氧类物质的产生，造成植物脂膜和叶绿素损坏，进而导致光合过程和植物整体生长的改变（Najeeb等，2017）。镉可以在不同的食用组织中积累（Baldantoni等，2016），导致根、茎和叶生长减慢，降低净光合作用和水分利用效率，并改变养分吸收（Rizwan等，2017）。

放射性核素也可能对食品质量构成潜在威胁，核能事故可以导致大气中的放射性核素沉积在土壤中，化肥、废料和核工业副产品也可以向土壤中引入放射性核素（Mortvedt, 1994）。放射性核素从土壤转移到植物和食物链的可能性，在20世纪50年代首次在核武器试验限制区以及炸弹试验产生的放射性沉降物中都得到了确认。1986年切尔诺贝利核事件对土壤造成了广泛的放射性核素污染（主要是^{131}I、^{134}Cs和^{137}Cs）（Bell、Minski和Grogan，1988）。土壤吸收污染物并进入牧草之后，会使得食草动物也受到污染，这导致了大不列颠及北爱尔兰联合王国对来自受污染地区的羊的销售和屠宰实施了限制（Smith等，2000）。在日本福岛核事件之后，放射性核素也出现了对食物链的普遍污染（Berends和Kobayashi，2012）。

与金属、类金属和放射性核素相比，受有机污染物严重污染的土壤在全球分布要小得多，相关的食物链污染主要集中在工业或城市中心周围。污染是由于废物在土地上的再利用，以及持久性和可在生物体中积累的有机化学品残留或废弃处置造成的，典型的化学品有有机氯、二噁英、多氯代二苯并呋喃以及全氟和多氟烷基物质。由有机物导致的土壤污染的程度通常低于金属/类金属，二噁英、呋喃以及全氟和多氟烷基物质等有机物在土壤中的浓度数量级一般小于毫克/千克。

有机污染物通过吸收进入食物链的途径取决于有机污染物自身的特性，主要包括挥发性、疏水性和水溶性。低挥发性的亲水性有机污染物（如全氟和多氟烷基物质）主要通过根部吸收进入食物链并转移到食物部分（Navarro等，2017）。挥发性疏水有机污染物（如二噁英、呋喃和多氯联苯）通常在土壤中具有强吸附性，往往通过大气吸收而在食物链中得到积累（Collins、Fryer和Grosso，2006；Simonich和Hites，1995）。有些植物也可以通过从土壤中吸收这些化合物，而使它们在植物体中得到积累（Huelster、Mueller和Marschner，1994）。相关研究利用^{14}C同位素标记技术，量化研究了有机污染及其在植物中的残留（Sun等，2018）。许多研究已经证实了农药残留可以通过多种主要途径被吸收。这些残留物进入植物组织并最终转移给消费者（Randhawa等，2014）。然而，与金属和类金属不同，有机物污染土壤对人类健康造成的不利影响的发生率和严重程度缺乏详细的记录或证明，这可能是因为受有机物污染的土地面积较小，且污染程度一般较低。

通过大气沉降和气孔进行气体交换是植物吸收持久性有机污染物的主要机制；这些污染物随后转移到其他植物组织中，并在它们的疏水性脂质和蜡质中积累（Odabasi等，2016）。通过植物根部从土壤中吸收的持久性有机污染物是有限的，因为持久性有机污染物被紧密结合在土壤颗粒上（Collins、Fryer和Grosso，2006）。因此，土壤可以作为持久性、中低挥发性有机污染物的储库和来源，这些物质可能被植物从大气中吸收后进入食物链。持久性有机污染物在胃肠道的吸收效率及其储存和释放动力学与脂肪的储存和代谢密切相关（Sweetman、Thomas和Jones，1999）。

2.2 农业土壤污染对生态系统服务的影响

据报道，污染物直接进入（在陆地上排放污水）或间接进入（使用受污染的水灌溉作物）土壤，污染了大片地区的土壤资源和地下水水体，通过食品污染影响了作物生产以及人类和动物的健康（Saha 等，2017）。

农业输入（如化肥、农药、动物粪便中含有的抗生素或植物中用于预防和治疗疾病的药物），是农业土壤的主要潜在污染物，并且新产品的研发使得其化学组成快速变化，其所带来的挑战具有特殊性[全球土壤伙伴计划（GSP），2017]。为生产足够食物、纤维和生物燃料而进行的农业集约化生产导致了土壤污染的延续。在中国，重金属含量在过去30年里大幅增加，与1990年的背景浓度相比，锌的含量增加了48%，镉的含量增加了250%（Zeng，Li 和 Mei，2008）。然而，污染物从土壤向植物的转移过程并没有被充分了解，"在更健康的土壤中生产的食物是否也更有营养？"需要更有力的科学证据来证明，以促使政策制定者、政府和土地使用者采取可持续以及环境友好的做法，并摒弃更多商业导向的做法。

2.2.1 化学肥料

为了提高生产力并减少作物损失，现代农业中大量使用化肥和农药，这种操作加速了土壤污染。当污染物在土壤中达到较高浓度时，不仅会发生土壤退化过程，而且会影响到作物产量。因此，土壤污染除了危害人类健康和环境外，还会造成经济损失。

土壤中过量的氮可通过硝化作用和其他氮转化过程造成土壤酸化和盐渍化。在自然条件下，土壤酸化非常缓慢，可达数百万年到几亿年（Guo 等，2010），但农业操作（主要是过量的施用氮肥）显著加速了土壤酸化过程，在不同土地利用类型中土壤pH平均降低达0.26（Lucas 等，2011；Tian 等，2015；Zhao 等，2014a）。在对中国农业土壤酸化来源的分析中，相关研究表明来源于氮肥的人为酸化是中国农业土壤酸化的主要原因，比酸沉降引起的酸化高 10～100 倍（Guo 等，2010）。

2.2.2 酸化和作物损失

农业土壤的酸化可以通过有毒重金属的活化而进一步造成土壤污染。如果施用于农业土壤的氮含量高于植物自身的需求，微生物硝化过程将导致硝酸盐（NO_3^-）的积累，这些硝酸盐由于溶解度高，会很容易渗透到地下水中，从而污染地下水（Tian 等，2015）。当土壤可用养分增加时，微生物量和活性也随之增加，造成微生物多样性被改变，从而导致养分循环不均衡（Lu 和 Tian，2017）。

磷肥的主要风险是其在被转移到表层水体中时会在许多地区造成水生生态系统的富营养化（Stork 和 Lyons，2012；Syers，Johnson 和 Curtin，2008）。磷元素会通过吸附于被侵蚀土壤颗粒、过量施用磷肥、在条件不适宜的情况下使用粪肥等转移到水体中（Syers，Johnson 和 Curtin，2008）。许多农田施用的磷量远大于作物能够吸收的量，这至少会在短期内导致土壤磷过剩（Aarts，Habekotté 和 Keulen，2000；Syers，Johnson 和 Curtin，2008）。

2.2.3 农药

政府间土壤技术小组最近就关于农药使用对土壤功能影响的相关研究进行了广泛回顾（FAO 和 ITPS，2017）。该工作提供的主要科学证据表明，当农民使用农药时，他们的净收益增加了，但是这一结论是建立在使用合成农药与不使用农药之间的比较上的，而不是将使用合成农药与使用生物防治害虫方法相比较（Cai，2008）。

也有报告指出，一些特定农药对土壤生物和土壤功能存在负相关影响。例如，一些有机氯农药会抑制共生固氮，从而导致作物减产（Fabra,1997；Fox 等，2007；Santos 和 Flores，1995）。联合国粮农组织和政府间土壤技术小组的报告还指出了当前农药与土壤健康间相互关系仍存在许多知识空白，尤其是在土壤污染方面（FAO 和 ITPS，2017）。经过国际努力，《鹿特丹公约》和《斯德哥尔摩公约》通过了关于评估农药的生态毒理学风险并控制其在环境中使用和释放的条约（UNEP，1998，2001），这是预防和控制土壤污染的重要成果。但是，关于它们与土壤组分的具体相互作用、它们在土壤介质中的流动性、可能的植物吸收以及它们对作物产量的影响，还需要更多的相关信息（Arias-Estévez 等，2008）。特别是对中低收入的国家而言，在市场上可以购买的大量农药中，并不是每一种化合物在获得授权之前都对其生态毒理学效应进行了分析（Aktar，Sengupta 和 Chowdhury，2009）。一些因对人类健康和环境存在严重不利影响而在高收入国家停止市场销售的农药，仍然可以在中低收入国家继续注册使用。由于农药残留在整个生态系统中都存在，制定关于土壤、地表水、地下水以及饮用水，特别是食品中的农药残留水平监测方案非常重要。然而，在许多低收入和中等收入国家，由于缺乏监管能力，相关监管方案基本不存在（Brodesser 等，2006）。

2.2.4 粪肥

施用未经处理的有机粪肥会造成重金属污染，不仅会对植物品质和产量的各项指标造成不利影响，还会导致微生物群落的大小、组成和活性发生变化，从而影响养分循环，降低养分有效性（Yao, Xu 和 Huang, 2003）。

如前所述，给牲畜服用的抗生素有很大一部分在动物的肠道内不能被吸收，会通过尿液和粪便排出体外。因此，未经处理的粪便会含有大量的抗生素，这可能导致土壤中抗生素耐药性的迅速增加（见2.3.2节）。在过去几年里，土壤中抗生素的命运和影响受到了极大关注，O'Neill委员会的报告显示，到2050年抗生素耐药性感染可能成为世界上导致死亡的主要原因。

随粪便和排泄物进入土壤的最常见肠道病原体是沙门氏菌、弯曲杆菌和大肠杆菌。在施用于土壤之前，病原体浓度水平随贮藏期的时间延长及温度升高而降低（Garcia 等，2010）。一旦在土壤中传播，病原体可以存活数月或数年。

2.2.5 农业生产中的城市废弃物

将污水污泥应用于改良土壤有许多积极作用，如减少废弃物数量、促进养分循环、增加土壤肥力、改善土壤结构和土壤保水能力等，这些比其所产生的负面作用更为重要。该方面工作的重点主要应该放在减少污水污泥以及用于灌溉的废水中的污染物浓度上。正如Petrie等人所强调的，在对新型污染物以及废水和污水污泥中存在的其他污染物的环境行为缺乏全面了解的情况下，在其应用到土壤之前必须要进行严格的分析（Petrie, Barden 和 Kasprzyk-Hordern，2015）。

在作为改良剂应用于土壤之前，堆肥和预处理可以减少城市垃圾中污染物和病原体的含量，并为废弃物熟化和将其转化为有价值的有机肥料提供了一种经济和环境友好的方法。然而，高浓度的重金属如Pb、Cd、Cu、Zn、

Cr、Ni和盐类仍然会存在于这种改良剂中，可能会影响土壤性质并抑制植物生长（Bolan 等，2014；Hargreaves, Adl 和 Warman, 2008；Stasinos 和 Zabetakis, 2013；Stratton, Barker 和 Rechcigl, 1995）。在施用或掺入土壤之前，需要对不同污水处理厂产生的化学成分组成进行全面的调查（Bauman-Kaszubska 和 Sikorski, 2009；Bien, Neczaj 和 Milczarek, 2013）。与未经处理的污水污泥相比，经过堆肥处理的污泥中重金属的生物有效性和作物吸收都非常有限，这表明在其应用于土壤之前进行预处理是非常有必要的（Smith, 2009）。

2.3 与土壤污染有关的人类健康风险

Oliver和Gregory总结了6种与土壤有关的人类健康风险（Oliver 和 Gregory, 2015）。其中，3种与土壤污染有关：元素污染风险（如As、Cd、Pb）；有机化学污染（如多氯联苯、多环芳烃、持久性有机污染物）；药物污染（如雌激素、抗生素）。其他三种风险来自土壤病原体，分析为炭疽病毒和朊病毒、微量营养素缺乏、由土壤退化而导致的营养不足。

土壤污染对人类健康和环境的长期影响仍不清楚，目前正在尝试许多努力，以便更好地理解有毒污染物的自然衰减和对健康影响的机制（Bernhardt 和 Gysi, 2016）。城市土壤尤其值得关注，因为人类活动集中在这些土壤上，而且由于与营养、空气质量和获得预防疾病的卫生服务等其他有关健康的决定性因素的相互作用，相关的影响机制更为复杂（WHO, 2013）。然而，非城市地区也受到许多不同来源的污染，这些污染往往来自于扩散源，很难被追踪并估计它们的程度和风险。今后有关土壤污染控制和修复方面的工作，应努力将这些领域纳入其风险评估办法中。

2.3.1 人类接触土壤污染物的途径及其对人类健康的影响

前几章已广泛讨论了与工业、采矿、城市和农业土地利用有关的主要污染物。本节将重点讨论与人类健康最相关的土壤污染物及其风险。

人类接触土壤污染物的途径根据污染物本身以及特定地点的条件和活动不同而不同（Shayler, McBride 和 Harrison, 2009）。一般来说，人们可以通过摄入或食用已经积累了大量土壤污染物的食物或动物而接触到土壤中的污染物（Khan 等，2015）；或者在公园和花园等公共空间造成皮肤暴露（ChaparroLeal, Guney 和 Zagury, 2018）；又或者吸入挥发了的土壤污染物（图14）。人类也可能因供水和空气污染物沉降引起的二次污染而受到影响（西英格兰大学科学传播部，2013），在某些情况下，土壤作为污染物的来源在这两个过程中起着重要作用。

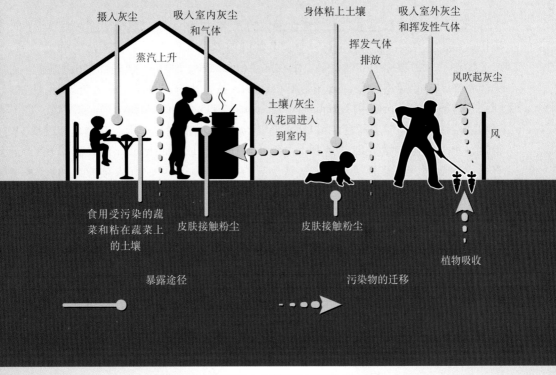

图14 在住宅环境中土壤污染可能的暴露途径
资料来源：EA, 2008。

根据世界卫生组织国际化学品安全计划，已经确定了十种或十组具有重大公共卫生问题的化学品（WHO, 2010）。这十种或十组化学品包括土壤污染物（如镉、铅和汞）、二噁英和二噁英类物质、高危险性农药（HHP）。这些污染物残留可以从受污染的土壤中转移到食物和水体中。HHPs的定义是根据国际公认的分类体系或在相关有约束力的国际协定或公约中列出的对健康或环境具有特别高的急性或慢性危害的农药。当在一个国家使用时会对健康或环境造成严重或不可逆转损害的农药被视为高危险物质（FAO 和 WHO, 2016）。所有农药的长期接触都会产生慢性影响，尤其是儿童、孕妇和营养不良的人，更容易受到农药接触的影响。与此同时，土壤中的病原体也可能污染食物，危害人类健康。从腹泻到癌症，有超过200多种疾病与摄入受污染的食物有关（WHO, 2017b），世界上有24%的人口受到土壤传播寄生虫感染，导致营养失衡和慢性贫血（WHO, 2017a）。

植物从土壤中吸收金属可能会对健康造成重大危害（Brevik, 2013；Burgess, 2013；Jordão 等, 2006）。植物根系对重金属的吸收是重金属进入食物链的主要途径之一，其吸收程度因植物根系的不同而存在差异（Pan 等, 2010；

Wagner，1993）。镉和铅是对人类毒性最大的元素（Volpe 等，2009）。食物是人类摄入镉的主要来源。日本的一个著名案例便是，人们食用了被镉污染的大米，导致了一种被称为"痛痛病"的疾病（Abrahams，2002）。通过食物摄取的镉能在怀孕期间穿透胎盘，破坏胎膜和DNA，扰乱内分泌系统，并可诱发肾、肝和骨骼损伤（Brzóska 和 Moniuszko-Jakoniuk，2005；Souza Arroyo 等，2012）。铅的毒性作用影响多个器官，可导致肝、肾、脾、肺的生化失衡，并引起神经毒性，主要发生在婴儿和儿童身上（Guerra 等，2012；Jaishankar 等，2014）。有机汞化合物，尤其是甲基汞，被认为是有剧毒的。汞可引起人体神经和胃系统的病变，并可导致死亡。砷可经口或呼吸道被人体吸收，主要储存在肝、肾、心脏和肺中，少量积聚在肌肉和神经组织，并被定义为致癌物（Brevik，2013）。它会导致神经系统紊乱、肝肾衰竭以及贫血和皮肤癌。镍会引起胃、肝、肾缺陷并影响神经系统（Brevik，2013）。锌与贫血和组织病变有关。虽然铜对健康的影响比较少，但如果长期接触，可能对婴儿的肝脏和肾脏造成损害（Brevik，2013）。

已经有越来越多的人意识到蔬菜和水果对人类饮食的重要性，并确认食品是许多污染物的重要来源，这表明应该定期监测农作物中的重金属污染情况。世界卫生组织和联合国粮农组织制定了《国际食品法典标准》（WHO 和 FAO，1995），确定了水果、蔬菜、鱼类和渔业产品以及动物饲料中污染物的安全限值。

Aktar, Sengupta 和 Chowdhury 对欧盟食品中农药残留污染情况进行了综述。尽管食物中的农药残留量没有超过日均可接受摄入量，但有少量研究已经分析了与生物体中持久性污染物有关的长期风险（Kim, Kabir 和 Jahan，2017；Xu 等，2017）。Hernández 等人强调需要进一步研究农药混合物对人类健康的影响，因为目前的立法考虑的是食物和水中单一农药的最大残留水平，而没有考虑低浓度时各种农药间可能的相互协同作用（Hernández 等，2013）。职业性接触农药与各种疾病有关，包括癌症、激素紊乱、哮喘、过敏和过敏反应（Burgess，2013；Van Maele-Fabry 等，2010）。

土壤中积累的持久性有机污染物的摄入与人类健康有较高的相关性（图15）。世界卫生组织/联合国环境规划署全球监测计划的最新结果显示，在世界上许多地区，母乳中的二噁英、呋喃和多氯联苯的水平仍远远高于毒理学上认为安全的水平，尤其是在印度以及一些欧洲和非洲国家，发病率较高（van den Berg 等，2017）。在许多非洲和南美洲国家，儿童、孕妇和哺乳期妇女吞食土壤（食土癖）是一种常见的现象，这种现象已扩展至西方社会（Reeuwijk 等，2013）。摄入含有持久性有机污染物和重金属的黏土是土壤传播疾病的重要来源，因为这种情况下所摄入的污染物经常超过安全日常接触水平（Odongo, Moturi 和 Mbuthia, 2016）。Bányiová 等人指出在捷克共和国持久性有机污染物暴露的主要来源是摄入受污染的食品（Bányiová 等，2017）。尽管自实施《斯德哥尔摩公约》以来，人体中的持久性有机污染物水平已有所降低，但相关事件仍然时有发生，持久性有机污染物仍是土壤和食品污染的一个重要来源（Hilscherova 等，2007）。因为持久性有机污染物在母体内是循环流动的，母乳中持久性有机污染物的存在会对新生儿和胎儿的健康构成非常高的风险（Reeuwijk 等，2013）。

摄入受污染的食物是多环芳烃危害人体的主要途径，它们具有潜在的致癌风险（Brody 等，2007；Xia 等，2010）。由于多环芳烃的芳香特性，它们很容易穿透细胞膜并与DNA分子共价结合，从而导致突变（Muñoz 和 Albores, 2011）。多环芳烃的暴露和毒性存在许多不确定因素，因此建立其健康风险评估是非常复杂的。

关于新型污染物，尽管我们已经获得了有关人体组织中暴露途径和水平的信息，但是我们在理解它们在环境中如何表现、在土壤基质中发生哪些相互作用、它们在人体中的毒性、生物积累特性以及它们在人体内的转移机制方面仍然存在知识空白（Covaci 等，2011）。通常情况下，这些化合物在人体中的浓度非常低，并且其中很多化合物只是最近才被认定为污染物，关于其流行病学方面的作用机制仍需要进行长期的研究。

当前，仍缺乏相关研究来评估大多数土壤、重金属和化学污染对健康的综合影响（Landrigan 等，2018）。为了分析土壤污染物对人类健康的影响，需要基本毒理学数据以及有关接触途径和等级的知识。毒理学中的可容忍剂量通常被用来进行污染物危险剂量的风险评估（Blume 等，2016；WHO，2013）。这些可容忍剂量被用来在确定来源、途径和受体接触情景下，说明某种物质对人

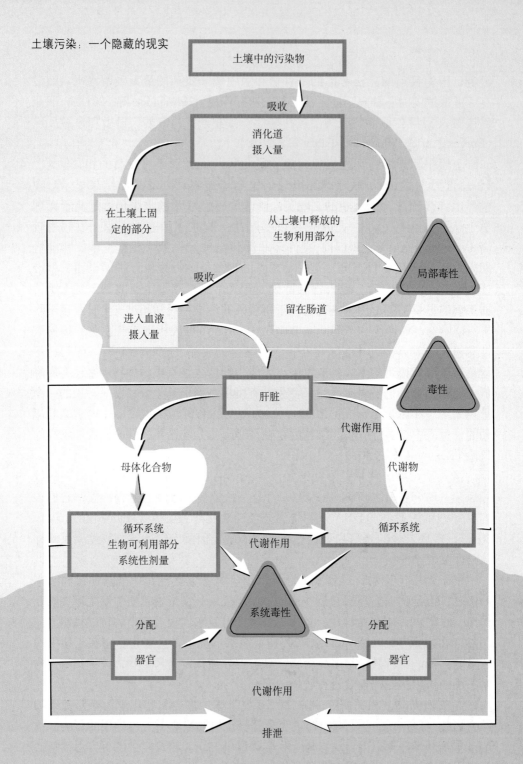

图15 口腔摄取土壤污染物对人体危害途径
资料来源：Hosford, 2008。

体健康的有害影响。

正如世界卫生组织报告所述（WHO，2013），有几种科学工具可用于评估土壤污染对人类健康构成的风险。最近开发的其他评估人体暴露的模型包括BROWSE（用于旁观者、居民、操作人员和工人暴露的植物保护产品模型），该模型包含了更现实的场景（ButlerEllis 等，2017），并为了完善人体暴露的评估，整合了大量欧洲的指导和监管数据库（Lammoglia 等，2017）。在现实环境中同时接触两种或两种以上化学物质是非常常见的，它们可能产生协同效应，但健康风险评估对此尚未有明确定义（Nicolopoulou-Stamati 等，2016）。

2.3.2　土壤是抗生素耐药性细菌和基因的储存库

随着抗生素整体有效性的下降，抗生素耐药基因从环境向人类的转移成为一个巨大的挑战（Harbarth 等，2015；Thomas 和 Nielsen，2005；WHO，2018）。这种感染会在体内持续存在，增加了感染他人的风险（WHO，2018）。每年全球约有70万人死于耐药性细菌，其中欧洲约25 000人（EC，2017），美国约23 000人[美国疾病控制与预防中心（CDC），2013]。此外，尽管研究人员已经观察到一些潜在的不利影响，但是摄入食物中残留的抗生素和抗生素耐药性细菌对人类健康的影响在很大程度上仍是未知的。

这些包括过敏和中毒反应或长期低剂量接触所造成的慢性中毒效应（McManus 等，2002；Sarmah，Meyer 和 Boxall，2006）。抗微生物药物耐药性的风险在新生儿中尤其重要，因为AMR细菌遍布新生儿的内脏（Brinkac 等，2017）。

土壤被认为是携带一系列抗药性细菌的天然宿主，这些抗药性细菌携带着许多已知和未知的耐药性基因（Cytryn，2013）。天然存在于环境中的真菌和细菌产生了许多人类使用了几个世纪的抗生素，同时它们也拥有对抗它们产生的抗生素的抗药性基因（Hopwood，2007）。由粪肥或污水污泥引入到土壤中的外来抗性细菌和基因可能不能很好地适应土壤条件，因为它们会受到土壤中原有生物体的选择压力（Heuer 等，2008）。

当微生物（如细菌、真菌、病毒和寄生虫）持续暴露于可以杀死或抑制微生物生长的抗生素或其他抗菌剂时，就会出现耐药微生物的选择，即使在低浓度情况下也是如此。同样，耐药基因组（染色体外抗生素耐药质粒）或染色体内的突变基因也可以转移到同一菌种和其他菌种中（Khachatourians，1998）。

抗生素在世界范围内被广泛用于动物治疗和牧业生产中的促生长。这些抗生素有很大一部分不能被生物体吸收，而是排泄到环境中。家畜的同化量取决于动物的抗生素代谢动力和动物新陈代谢对药物的转化能力。例如，Heuer 等人发现，在给猪用药10天后，96%以上的兽用抗生素磺胺嘧啶以亲本形式或代谢物的形式排出体外（Heuer 等，2008）。四环素类的排泄率较低（Winckler 和 Grafe，2001），但阿莫西林和双氟沙辛的排泄率可达到90%以上（Sukul 等，2009）。在废水处理或堆肥过程中，抗生素并不能被完全清除，这些抗生素最终大量地进入农场和城市废物以及土壤中。污水处理厂被确定为向河流中释放抗生素的主要来源，且不同抗生素的去除率差异很大（Michael 等，2013；Watkinson 等，2009）。

在土壤中施用改良剂后，土壤微生物群落及其主要活性发生了变化，促进了抗性种群的形成（Ding 等，2014；Tian 等，2015；Tien等，2017）。粪便中的抗菌素耐药群体，可能是耐药质粒向土壤微生物转移的重要原因，这一猜测在施用粪肥的土壤中得到了证实（Gotz 和 Smalla，1997；Smit 等，1991）。这一转移过程通常随着营养来源的增加而被促进，与此同时，微生物活性和种群密度被提高（Ding 等，2014）。科学证据表明，土壤中的重金属，尤其是铜和锌，有助于抗生素耐药性的协同选择（Grass, Rensing 和 Solioz，2011；Hölzel 等，2012；Wales 和 Davies，2015；Yu 等，2017）。土壤微生物发生的突变可能导致一系列代谢表型，包括利用不同碳源、氮源或磷源能力的变化（Perkins 和 Nicholson，2008），进而对全球地球化学循环形成影响（Allen 等，2010）。

Wales 和 Davis 还发现，即使在没有抗菌物质的情况下，当暴露于其他杀菌剂（如消毒剂和防腐剂）时，重金属也会增强对抗菌物质的选择。在澳大利亚居民区的土壤中发现，高浓度重金属刺激了抗生素抗性基因的增殖（Knapp 等，2017）。因此，抗生素耐药性在受污染土壤中面临更大的挑战。对此，需要对抗生素耐药性通过污水污泥和粪肥改良传播的风险进行更多地研究（Bondarczuk, Markowicz 和 Piotrowska-Seget，2016）。

抗生素通过不同过程进入土壤后,通过植物吸收、淋溶到地下水中,与有机物发生络合被固定、被黏土矿物吸附或生物降解等过程,其浓度会迅速下降(图16)(Jechalke等,2014;Kuppusamy等,2018)。尽管如此,即使在抗生素浓度很低的情况下,也能观察到耐药性细菌的增加(Gullberg等,2011)。抗生素在土壤中的固持和吸附降低了其生物有效性,但增加了其在环境中的持久性(Jechalke等,2014)。由于食物链可能直接促进抗生素耐药性的传播,因此植物对抗生素的吸收已被广泛报道和关注(Boxall等,2006;Du和Liu,2012)。抗生素也可能会抑制种子萌发,降低作物生长(Du和Liu,2012)。此外,在富含抗生素的土壤中,有机物的浓度和结构变化会导致被隔离的抗生素以生物可利用的形式被释放出来(Gulkowska等,2013;Rosendahl等,2011)。

图16 兽用抗生素在环境中的去向
资料来源:Kuppusamy等,2018。

土壤污染：一个隐藏的现实

不同的抗生素在细胞内有不同的靶点，因此生物体产生了对特定抗生素的耐药性，而不是普遍的耐药性（Khachatourians, 1998）。然而，细菌对多种药物产生耐药性的现象越来越普遍（CDC, 2013；EC, 2017；WHO, 2014）。最近的一项研究表明，从鸡粪中分离出的多药耐药菌群落中存在抗生素耐药基因和可移动遗传因子的高度共存（Yang等, 2017b）。细菌产生耐药性存在四种主要机制（图17）：①细胞内和细胞外的细胞酶对抗生素化合物的酶催化降解或修复；②外排泵能主动清除细胞外或细胞质内的抗生素化合物；③抗生素结合位点的修复或保护；④天然的或修复的膜渗透性（Alekshun和Levy, 2007）。

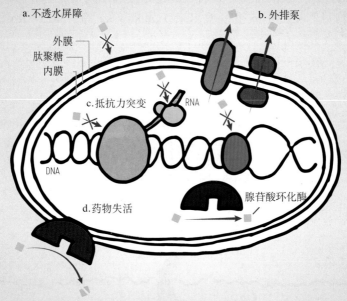

图17　抗菌素耐药性的主要机制
资料来源：Allen等, 2010。

3 污染土壤的管理和修复

对污染土壤进行评价和管理的第一步是鉴别问题，也就是确定土壤中的污染物。一般而言，当某一地区受到石油泄漏、核事故或坝体尾矿破裂等事故影响时，会立即采取措施控制其范围，并防止事故的进一步发生。然而，在有遗留污染的土壤中或存在扩散污染的地方，如何采取相关措施却成为一个难题，因为这往往没有既定的规程可遵循。在世界上，有的国家、区域或地方机构负责发起初步调查，以确定是否存在污染，是否需要采取进一步行动，然而还有许多地方并没有相关法规或协议（Teh 等，2016）。

在过去，土地复垦标准是根据背景浓度和安全限度标准制定的。新方法试图就污染物对环境、人类和食品安全构成的风险进行更全面的评估。因为土壤的复杂性，以及关于土壤中污染物生命周期、相关毒理学和综合研究资料的缺乏，完成污染物对环境和人类健康潜在风险的特性描述任务难度非常大（Cachada 等，2016）。模拟相关污染物的暴露路线时，通常会考虑特定土地的使用类型（如住宅、工业和娱乐）（Provoost, Cornelis 和 Swartjes, 2006）。

3.1 风险评估方法

风险评估是指，人们基于科学证据，评估某种结果的可能性和严重性，并利用这些知识来协助做出决策。评估中必须尽可能减少不确定性，并明确剩余的不确定性中需要清楚识别和解释的（FAO，2000）。关于土壤或沉积物风险管理决策的重点是确定对人类健康或环境构成风险的相关暴露途径，并制定适当的修复措施。这些措施可能包括移除来源、切断路径，或两者双管齐下（土壤和沉积物中污染物生物有效性委员会，2002）。

全球的风险评估方法是相似的，包括通过一系列步骤来识别和评估外源或本底物质已经导致的或正在造成的土壤污染，并评估会对环境和人类健康构成多大程度的污染风险（图18）。风险评估是基于科学制定相关政策和技术措施，并在必要时采取相关行动的重要工具。尽管综合方法正在变得越来越受欢迎，但主要的风险评估方式仍是通过一步步的化学分析来进行，集中于单一介质、单一来源和单一毒性终点。这种方法基于确定性或概率性分析，使用的模型结合了人体暴露和环境影响参数（DEA，2010；Hope，2006；

Provoost, Cornelis 和 Swartjes, 2006)。终端用户感兴趣的主要是可能存在局地污染和扩散污染的工业区和城市区的土壤是否"适合使用"。在这些情况下，需要一种针对特定地点的方法，以获得暴露和风险信息的综合概述(Posthuma 等，2008)。

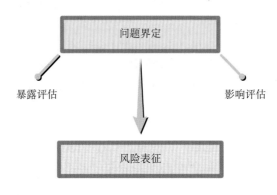

图18　通用风险评估模式
资料来源：Posthuma 等，2008。

一旦怀疑存在污染，并对场地的历史使用情况进行初步研究后，应进行初步评估，确定是否存在外源物质，存在哪些外源物质，以及它们是否对环境和人类健康构成风险。如果确认存在污染并需要采取修复措施，则必须完成详细地调查，以确定污染程度和可能采取的修复措施。随后应确定和实施风险管理或修复策略，并在完成后进行事后清理检查，以确认风险是否被降低，污染源是否已得到控制。

在世界范围内，政策和监管都是基于风险评估方法来预测那些无法直接测量的风险（Hough, 2007）。法规包括使用土壤质量标准来识别和评估土壤污染，在许多情况下需要考虑国家土壤特性或特定地点的条件。然而，由于风险评估方法是一个复杂且耗时的过程，世界上并不是每个国家都能负担得起污染调查。也因为没有全面的资料可用，所以经常采用站点研究的方法。正如Hope所指出的，获取有关生态风险评估及其监管依据的文件是复杂的，尤其是在发展中国家（Hope, 2006）。虽然各国在气候、土壤类型和当地人群的特点方面不尽相同，但美国环境保护署（US EPA,1986年）、加拿大指南[加拿大环境部长理事会（Canadian Council of Ministers of the Environment），1999]和荷兰指南（Brand, Otte 和 Lijzen, 2007）等一些资料仍可作为参考使用（Li等，2014）。一些国际组织，如联合国粮农组织（FAO, 2000）也提出了一些指导方针，如其提供了评估废弃农药库存所构成的环境和人类健康风险的相关指导，并在环境管理工具包中提供了更详细的评估步骤信息。几个国际组织制定的综合风险评估指南[国际原子能机构（IAEA），1998；Meek 等，2011；WHO, 2001a]也在试图提供一种综合的多化学、多媒介、多途径和多物种的暴露分析。

人们普遍认为，开发一种包含多种复杂污染物的风险评估工具是十分必要的，它能更好地了解污染物的潜在影响及其影响程度（Reeves 等，2001）。Albert在30多年前提出了这样一个问题："能否预测复杂混合物的毒性？"（Albert，1987）。从那时起，许多研究人员试图提出一种合适的解决方案，或至少是对复杂混合物的相互作用进行更全面地研究，以确定在多种混合污染物共存时，是否存在附加的、协同的或拮抗的毒性效应（Chen 等，2015）。由于工业操作或工艺实施的千差万别，每个地点的混合污染物都存在巨大的特殊性和差异性，这减缓了一般风险评估方法确定相关限值步骤的工作进展（Callahan 和 Sexton, 2007）。荷兰提出的方法中包括一项协议，可用于分析多种物质共存时的风险（Cachada等，2016）。其通常使用累积计算方法，考虑了个体风险和潜在毒性及风险的总和，而不考虑物质之间可能的相互作用和协同作用，这有可能降低或增加其潜在风险（Callahan 和 Sexton, 2007）。Chen 等人发现混合污染物越复杂，协同毒性越强（Chen 等，2015）。他们认为，与传统的浓度加和或独立作用模型相比，使用复合指数评估生态毒理学风险更准确（图19），这不仅适用于水生环境（Rosal 等，2010），也可以在土壤中使用

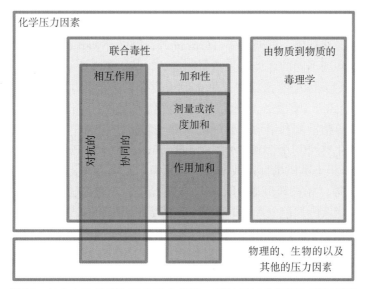

图19　包含由物质到物质的独立作用、剂量累加效应和物质间相互作用的风险评估方法

资料来源：Sarigiannis 和 Hansen, 2012。

(González-Naranjo 和 Boltes，2014；González-Naranjo 等，2015）。不仅是农药组合中（Yang 等，2017a），在其他复杂混合物中，如在垃圾填埋场（Baderna 等，2011）或铁路轨道中发现的污染物混合物中，也存在协同/拮抗作用。在最近的研究结果中，Wierzbicka 等人发现虽然每种污染物的单一浓度都不超过允许值，但是混合污染物对来自不同营养级的多种试验生物具有高毒性作用（Wierzbicka, Bemowska-Kałabun 和 Gworek，2015）。然而，正如 Sarigiannis 和 Hansen 所解释的，综合毒理学方法只适用于特定条件，相关的数据无法推广（Sarigiannis 和 Hansen，2012）。

上述处理受污染场所的一系列步骤只是一般步骤，而对于特定的国家或地区，通常可以省略或增加一些步骤[污染场地管理工作小组，1999；DEA，2010；联邦环境署 (FOEN)，2013；Luque，2014]。

为实现不同的目标，人类健康风险评估可以以不同的方法进行。这些方法可用于以下目的：

- 土壤质量标准的制定
- 特定场地的风险评估
- 制定治理目标
- 按干预优先级别对污染场地进行排序

土壤筛选值是基于一般暴露途径和情景（如居民区或工业区吸入挥发性气体）的一般土壤质量标准，许多国家采用这些标准来规范污染土壤的管理。土壤筛选值或土壤质量标准在世界各地有不同的术语：触发值、参考值、目标值、干预值、清理值、截断值等（Carlon 等，2007；Swartjes 等，2012）。此外，阈值是根据各国不同的环境政策战略制定的，基本不考虑土壤性质。

在被重金属污染的土壤中，总金属浓度提供的潜在风险信息很少（Naidu 等，2015）。区分识别重金属形态的植物可利用性对于防止土壤的管理方式使得重金属从植物不可利用形态变为可利用形态非常重要，可以通过生物试验来确定金属的生物可利用性和毒性（Romero-Freire, Martin Peinado 和 van Gestel，2015）。在修正土壤质量标准或阈值时必须考虑土壤性质，如 pH、土壤质地和有机质含量等，相关研究表明，在许多情况下，不考虑土壤性质的质量标准低估或高估了实际风险（Appel 和 Ma，2002；Bradl，2004；Rodrigues 等，2012；Romero-Freire, Martin Peinado 和 van Gestel，2015）。此外，在风险评估中分析并纳入生物有效性，而不是假设目标污染物是 100% 可被生物利用，会进一步优化修复工作，从而提高修复工作的收益（Naidu 等，2015；Romero-Freire, Martin Peinado 和 van Gestel，2015）。

在食品质量认证中，根据食品中重金属的含量来制定相应的法律法规是至关重要的。在国际发表的文献中，提出了确定土壤中允许重金属阈值的多种方法和评价标准（表7），为相关标准的建立提供了重要参考（Muñiz,2008）。许多国家都确立了以重金属含量为标准的土壤质量参考值，它们制定了各自的土壤保护和食品安全保障环境政策，其中由美国环境保护署（USEPA,1998,2014a）制定的标准尤其重要，该标准被世界上其他国家广泛遵循。这些标准以风险评估政策为基础，明确了背景浓度以及人类和环境毒理学的相关研究方法。就食品而言，联合国粮农组织的食品法典标准至关重要。它确定了食品中允许的污染物和毒素（包括重金属）的限值，并不断审查和更新（WHO和FAO，1995）。

土壤污染：一个隐藏的现实

表7 各国住宅用地中某些重金属的阈值

污染物（毫克/千克，以干重计）	比利时[1]	法国	德国[2]	大不列颠+植物[3]	大不列颠-植物[4]	匈牙利[5]	荷兰	波兰[6]	西班牙[7]	瑞典	澳大利亚[8] 一住宅+花园	澳大利亚[8] 一住宅+无土壤暴露	加拿大	中国	挪威	瑞士	美国
砷	110	37	50	20	20	15	55	2	—	15	100	500	12	30[9]	2	无	22
镉	6	20	20	30	8[10]	1	12	4	1～3	0.4	20	140	10	0.43	3	20	37
三价铬	300	130[11]	400	200	130	75	380	—	100～150	120	100[12]	500[12]	64	58.9	25	无	100 000
铜	400	190	无	无	无	30	190	150	50～210	100	7 000	30 000	63	31.7	100	1 000[13]	3 100
汞	15	7	20	8	8	0.5	10	—	1～1.5	1	200	600	6.6	—	1	无	23[14]
铅	700	400	400	450	450	100	530	100	50～300	80	300[15]	1 200[13]	140	37.5	60	1 000[16]	400
镍	470	140	140	75	50	40	210	—	30～112	35	400	900	50	27.5	50	无	1 600
锌	1 000	9 000	无	无	无	200	720	300	150～450	350	8 000	60 000	200	117.7	100	2 000[17]	23 000

注：修改自Provoost, Cornelis 和 Swartjes, 2006。
1 2002年7月8日颁布的土壤修复法令。
2 为土壤—人类暴露途径制定的国家立法标准。
3 有蔬菜园的居民区。
4 无蔬菜园的居民区。
5 匈牙利政府条例第10/2000号。
6 波兰表层土壤（0～30厘米）的土壤质量标准，为2002年土地利用（农业用地、森林、居民区）条例B组制定。
7 1990年10月29日关于在农业中使用污泥的第1310/1990号敕令（B.O.E.第262号，1990年11月1日）。pH为低于或高于7的土壤。
8 土壤和地下水调查水平指南。2011年《国家环境保护》（场地污染评价）办法》。
9 环境保护总局（1995）土壤环境质量标准。国家环境保护局，GB15618—1995。
10 pH为6、7、8的土壤净化标准分别为1毫克/千克、2毫克/千克、8毫克/千克（以干重计）。本比较中采用8毫克/千克（以干重计）的净化标准。
11 总铬。
12 铬（Ⅵ）。
13 1 000、4为土壤净化标准的总含度和可溶性浓度。本比较中采用1 000毫克/千克（以干重计）的净化标准。
14 23、6.1分别用来描述氯化汞和有机汞。本次比较采用23毫克/千克（以干重计）的净化标准。
15 基于血铅模型的预测结果，考虑了50%口服生物利用度。

16　1 000、0.1为土壤净化标准的总浓度和可溶性浓度。本比较中采用1000毫克/千克（以干重计）的净化标准。

17　2 000、5为土壤净化标准的总浓度和可溶性浓度。本比较中采用2000毫克/千克（以干重计）的净化标准。

3.2 污染场地修复的主要技术

Nathanail将可持续修复称为"以安全、及时的方式消除和控制不可接受的风险，并使修复工作的整体环境、社会和经济效益最大化的修复"（Nathanail, 2011）。可持续管理需要综合使用现有的最佳可用技术，不仅是在修复过程本身，而是在整个过程中，包括风险评估和风险降低。最佳管理措施（BMPs）是在管理、文化和结构化措施的单独或组合作用下，被学术界或政府的研究人员确认是对减少环境损害最有效和最经济的方式（Cestti, Srivastava和Jung, 2003）。土壤修复通常会在不同地点采用不同方法，对于污染物、土壤性质、土地利用、所有权和责任制度以及技术和经济因素的每一种组合，需要采用差异化的技术或技术组合（Swartjes, 2011）。

修复技术可分为两大类：原地（现场）和非原地（将污染土壤移出现场进行处理）修复。可用的修复方案包括物理、化学和生物处理，这些方案为大多数土壤污染提供了潜在的技术解决方案（Scullion, 2006）。无论是原地还是非原地修复，对污染物作用的最终结果可分为降低浓度、在不降低浓度的情况下降低生物有效性、隔离在惰性基质中、抑制和去除（Pierzynski, Sims和Vance, 2005）。污染场地的管理是一种针对特定场地的方法，包括特征描述、风险评估和修复技术选择，其主要集中于局部或点源污染。

Scullion回顾了污染土壤修复的主要处理方法及其对污染物的影响（Scullion, 2006），详细说明了污染物是被降解、从土壤成分中分离、从基质中提取还是稳定存在的（表8）。

表8　主要修复方法及其对土壤污染物的影响[√＝主要过程，（√）＝受污染程度或范围限制的辅助过程]

工艺处理	破坏/降解	土壤分离	提取/损失	稳定
物理修复方法				
热力学	√		√	
凝固	(√)			
蒸汽提取			√	
空气喷射	(√)		√	

(续)

工艺处理	破坏/降解	土壤分离	提取/损失	稳定
清洗/抽水处理	(✓)		✓	
电修复法	(✓)		✓	
颗粒分选		✓		
化学修复方法				
氧化	✓		✓	✓
还原	(✓)		✓	✓
水解	✓		✓	
增溶	(✓)		✓	
脱氯	(✓)			
pH 处理	(✓)		✓	✓
生物修复方法				
微生物活性				
土耕	✓		(✓)	✓
生物打桩	✓		(✓)	
堆肥	✓		(✓)	✓
生物反应	✓			(✓)
生物淋滤			✓	
植物活性				
植物固定	(✓)		(✓)	
植物提取	(✓)		✓	(✓)
植物降解	✓		(✓)	(✓)

资料来源：Scullion, 2006。

  目前许多可用的物理方法成本高的部分原因是其需要挖掘和运输大量受污染的材料进行非原地的一些处理，如化学失活或热降解等。高昂的成本导致人们对原地应用替代技术越来越感兴趣，特别是那些基于植物和微生物的生物修复技术（Chaudhry 等，2005）。生物修复是一种利用特定微生物的生物活性，破坏或使各种污染物无害化的技术。生物修复实际上依赖于微生物的生长和活性，其有效性高度依赖于所应用环境中影响微生物生长和降解速率的环境参数。生物修复被认为是一项非常有前途的技术，在处理某些类型的污染场地时具有巨大的潜力（Zouboulis, Moussas 和 Nriagu, 2011）。生物修复已经在世界各地得到应用，并取得了不同程度的成功（Zouboulis, Moussas 和 Nriagu, 2011）。

Alexander认为，土壤中通过微生物活性进行的生物修复必须满足几个条件（Alexander，1999）：①微生物必须出现在含有农药的土壤中；②微生物必须有进行生物降解所必需的酶类；③农药必须能被具有必需酶类的微生物接触；④如果导致降解的初始酶是胞外酶，则酶类起催化作用的化学键必须暴露出来；⑤如果催化初始降解的酶类是细胞内的，那么分子必须穿过细胞表面到达酶作用的内部位点；⑥因为作用于许多合成化合物的细菌或真菌的种群或生物数量最初很小，所以土壤的条件必须有利于潜在活性微生物的增殖。

由锯末、木屑、树皮、稻草、植物废料和厨余垃圾制成的堆肥是另一个常见的向土壤中引入有机物的来源（Kuo等，2004）。向土壤中引入有机物可能有助于减少重金属和其他污染物的流动性（Grobelak和Napora，2015；Wuana和Okieimen，2011），从而降低其对环境和人类健康的风险。

添加粪肥和污水污泥是生物修复的有效工具，但要谨慎并确保有机物已经经过有效的预处理。为降低禽畜粪便对环境的负面影响，可在将其应用于土壤前，先进行一些简单的例如堆肥类的处理（Zhang等，2015a）。与新鲜有机肥相比，堆肥后有机肥的木质素和多酚含量普遍较高，这将会降低CH_4的排放，同时进一步提高有机碳封存的潜力（Xia，Wang和Yan，2014）。Lv等人观察到蚯蚓对堆肥过程具有积极作用，它可以稳定动物粪便中的重金属（Lv，Xing和Yang，2016）。新鲜肥料的堆肥已被证明是一种减少各种环境病原体和抗生素耐药细菌的有效方法（Cole，2015；Holman等，2016）。为了在施肥前降低粪肥和料浆中的病原体水平，Nicholson等人提出了将料浆储存一至三个月、进行高温堆肥以减少潜在挥发的方式施用、避免长途运输等一些建议（Nicholson等，2003）。相关研究观察到土壤中某些抗生素具有持久性，并且由于它们对土壤吸附性强，矿化极其微弱，因此，一些科学家强调了储存时间和堆肥的重要性，以便在肥料施用前消除粪肥中的抗生素化合物（Arikan，Mulbry和Rice，2009；Halling-Sørensen等，2001；Kim等，2011；Tien等，2017）。

种植对高浓度有毒物质具有良好抗性、对污染物具有较高收集和储存能力的树木也是土壤生物修复过程的良好措施（Paz-Alberto和Sigua，2013）。根据Wisłocka等人的研究结果，最普遍的具有重金属元素高富集能力的树木是黄桦（垂枝桦）、桤木（细叶桤木）、洋槐（刺槐）、柳树（柳属）和针叶树（Wislocka等，2006）。例如奇岗芒等能源作物对栖息地环境的变化有很好的适应能力，可逐步恢复退化土地，并有能力阻止重金属迁移到土壤和地下水中。

土壤污染：一个隐藏的现实

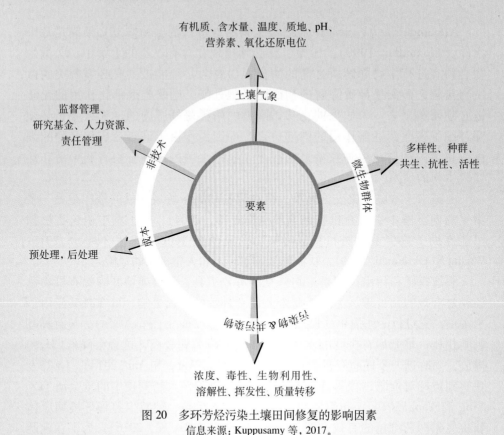

图20　多环芳烃污染土壤田间修复的影响因素
信息来源：Kuppusamy 等，2017。

当前，科研人员对生物炭表现出极大兴趣，尤其是关于生物炭颗粒的化学和物理特性如何影响土壤中的水分流动、去除污染物、改变微生物群落和减少温室气体的排放。生物炭有希望能够帮助世界各地的农民，特别是那些经常与贫瘠土壤作斗争的发展中地区的农民。生物炭历史起源较早，早在数百至数千年前，亚马孙的居民通过加热有机物质生产它，从而创造出富饶肥沃的土壤，称为普雷塔土壤（黑土）。大约在欧洲国家入侵南美洲的时候，这种做法就被抛弃了，而其他地方很少有农民会照常使用生物炭。科学家们第一次对这种材料产生兴趣是在大约十年前，全球变暖导致一些研究人员声称生物炭是一种将大量碳储存在地下的方法。由于生物炭的高成本，这种应用的可能性有所减弱，但土壤科学家现在正在探索其在农业和土壤污染修复方面的应用（Cernansky，2015）。

将纳米颗粒应用于受污染的土壤是一种新型的修复技术（Pan 和 Xing，2012）。在土壤修复中应用最为广泛的是纳米零价铁（nZVI），它可同时减少有机和无机污染物的影响。例如，nZVI可以有效降解氯代烃和有机氯农药（Singh等，2011；Zhanqiang，2010）。碳纳米管已被证明是一种可行的修复材料，因为它具有较大的重金属离子、放射性核素和有机物质吸附能力（Rao, Lu 和 Su, 2007；Ren 等, 2011；Pan 和 Xing, 2008）。

电动力学修复、酶介导的生物修复、多过程植物修复以及微生物修复等综合方法和新兴技术已经应用于多环芳烃污染土壤的修复中（Kuppusamy 等，2016）。最佳可用技术的选择及修复受污染土壤成功与否将取决于各物理、经济、管理和技术因素（图21）（Kuppusamy 等，2017）。

多氯联苯、多环芳香烃和多溴二苯醚等在土壤和沉积物上的强烈吸附严重影响着含有这类分子的土壤修复，尽管它们被严格限制使用已有30多年的时间，但是它们仍持久存在，这说明了其吸附作用之强。在大多数情况下，清除这些污染物的能力决定了修复技术的有效性（Gomes, Dias-Ferreira 和 Ribeiro, 2013）。对这些污染土壤最常用的修复技术是"挖掘和倾倒"，但这是不可持续的。生物修复、热解吸、厌氧脱氯等其他技术在近年来进行相关试验，并取得了良好效果（Gomes, Dias-Ferreira 和 Ribeiro, 2013）。上述技术虽然旨在破坏或转变多环芳烃，但操作方式差异很大，因此受不同的清理时间、成本、降解产物和环境影响。它们的有效性也因不同地点而有所差异，因为每项技术的应用效果都取决于污染物、污染物的老化程度、土壤类型和地貌条件，以及在环境中污染物的流动性或对土壤颗粒的吸附性等因素（Gomes, Dias-Ferreira 和 Ribeiro, 2013；Wang 和 He, 2013）。

3.3 降低食物链污染和生态系统服务的农业生产方式改变

可持续土壤管理自愿指南旨在为各国家、农民和其他利益相关方提供普遍接受的、经过实践证明的、基于科学的原理，以促进可持续土壤管理（FAO，2017）。这些准则对可持续土壤管理的描述如下："如果土壤所提供的支持、供应、调节和文化服务得到维持或加强，并且不显著损害能提供这些服务的土壤功能或生物多样性，那么土壤管理就是可持续的。"本章所引用的农艺措施有助于推动可持续土壤管理（FAO，2017）。

3.3.1 化肥

综合作物管理是一种平衡了营利性经营和环境敏感性之间关系的农业方法。它为农业面临的许多问题提出了一个现实的解决方案，其包括可用来避免浪费、提高能源效率和尽量减少污染的多种措施。综合作物管理结合了现代技术的优点和良好农业实践的一些基本原则，是一个面向全农业的长期战略（EC，2002）。

田间作物综合作物管理的组成部分如下：

1. 量化养分来源：土壤储存、粪肥、作物残余物；
2. 土壤试验：pH、石灰需求量、磷、钾（钙、镁可选）；
3. 粪肥分析：氮（铵态氮、总氮）、磷、钾；
4. 粪肥和肥料喷洒器的校准：吨、1 000加仑、磅每英亩；
5. 施肥计划：粪肥施用、追肥；将多余的粪肥用在替代作物（干草作物）上；使用土壤测试来避免因为大量施用粪肥导致磷含量过高；避免在粪肥或其他肥料施用时产生过量氮；以及基于土壤氮测试来侧施或追施氮肥；
6. 覆盖作物：减少土壤流失和硝酸盐淋失；在蔬菜田或者距离养殖场较远、不施用粪肥的农田中种植豆类覆盖作物；
7. 种植计划：确保及早收获作物，以便在大多数易受侵蚀的土地上及早种植覆盖作物；
8. 少耕：减少土壤侵蚀造成的养分流失。

土壤肥力综合管理主要基于以下原则：①仅以矿质肥料或有机质管理为基础的做法都不足以实现可持续农业生产；②具有良好适应性、抗病、抗虫害的种植资源是有效利用可用养分的必要条件；③良好的农艺措施，包括种植日期、种植密度和杂草管理等，对于确保有效利用稀缺营养资源至关重要[国际农业研究磋商组织和气候变化、农业和食品安全研究项目（CGIAR & CCAFS），2018]。此外，还需要确定作物轮作周期内的营养资源目标，最好包括豆科植物。

养分综合管理可以在促进植物生长方面发挥作用。干物质分配与作物总生物量（Amanullah 和 Inamullah，2016；Amanullah 等，2016），包括根系生物量（Amanullah 和 Stewart，2013），对退化土壤的植物修复过程的效率有显著影响（Grobelak，2016）。保护有机碳丰富的土壤、恢复和改善退化的农田、增加土壤碳含量，在解决粮食安全、使食物生产和人类适应气候变化，以及减少人为排放这三方面的挑战都发挥着重要作用[联合国气候变化框架公约（UNFCCC），2015]。

生物肥料是含有不同类型有益微生物活细胞的产品（细菌、真菌、原生动物、藻类和病毒），在土壤肥力、作物产量和盈利能力方面发挥着重要作用。农业中较为常用的有益微生物包括根瘤菌、菌根、偶螺菌、芽孢杆菌、假单胞菌、木霉菌和链霉菌。有益微生物对于分解土壤中的有机质以及增加农作物必需的大量营养素（氮、磷、钾、硫、钙和镁）和微量营养素（硼、铜、氯、铁、锰、钼和锌）是必不可少的。有益微生物在固体废物和污水管理中也发挥着重要作用。有益微生物可以提高植物对不同环境胁迫（如干旱、热、冷、盐）的耐受性，并增强植物对害虫和疾病的抵抗力。有益微生物不仅通过增加光合作用、产生激素和酶来促进作物生长和提高产量，而且通过控制不同的害虫和各种植物病害来提高作物质量。有益微生物减少了作物对化肥的需求，从而减少了由化肥引起的环境污染。它们降低了生产成本，从而增加了种植者的收入和盈利能力。有益微生物对提高作物产量、盈利能力和可持续发展具有重要意义。施用有机肥，如动物粪肥、家禽粪肥、绿肥、堆肥、农场粪肥、生物炭和灰，可以增加土壤中的有益微生物，提高土壤健康水平以及整体可持续性（Amanullah，2015）。

3.3.2 农药

为了实现一个无污染的世界，自愿性可持续土地管理指南在世界范围内被推荐（FAO，2017），其中包括综合管理方法和有机虫害管理方法。

害虫综合管理是一种基于预防、监测和控制的管理，其提供了消除或大幅减少杀虫剂使用的方法，从而降低农药对人类健康和环境的风险。害虫综合管理通过利用各种方法和技术来做到这一点，包括利用栽培、生物和结构策略来控制大量的害虫问题（Beyond Pesticides，2018）。此外，害虫综合管理鼓励使用作物轮作，这可以大大降低对农药的需求（García-Préchac 等，2004）。

在集约型农业生态系统中，最常见的农药施用方法为喷洒，其他农药施用方法，如种子处理、把颗粒撒在地面上、土壤灌药以及土壤烟熏消毒的使用较少。农药使用量的30%~50%会由于地面沉积、喷雾漂移到达邻近的环境区域或因挥发而损失，以至没有到达目标害虫（Diaconu 等，2017；Viret 等，2003）。"污染者付费"原则（将环境和公共卫生费用加在消费者支付的价格中）是将使用杀虫剂的社会成本内部化的有效方法。产生的费用和税收可用于促进可持续虫害管理（Popp，Pető 和 Nagy，2013）。控制滥用农药，并推广更环保的技术，如生物害虫防治（Popp，Pető 和 Nagy，2013），有助于减少农业领域的污染。

综合杂草管理是指通过长期的管理方法，采用物理防治、化学防治、生物防治和栽培防治等多种杂草管理技术来控制杂草的生长。

如前几节所指出，对食品质量产生不利影响的最普遍的土壤污染类型与金属、类金属和放射性核素有关。为了减少这些污染物引起的食物链污染，研究人员开展了关于相关农业措施的大量研究。由于有机化学品对土壤的污染在区域范围内一般非常有限，所以对这些化学品的研究就少得多，在这里也不作进一步考虑。

3.3.3 金属

在食物链污染方面，镉是研究最广泛的金属，有许多方法可以减少植物从土壤中吸收镉（Grant 等，1999）。表9对它们进行了总结，并可分为对作物（种类、品种和轮作）调控方法、土壤条件调控方法和灌溉水属性调控方法。

表9 减少镉对食物链污染的农艺管理措施

作物处理	土壤处理	水处理
植物种类	场地选择	使用低（氯）盐度的水
植物品种	耕作（稀释/埋葬）	
作物轮作	添加石灰	
植物提取	添加锌	
	添加吸附剂	

40多年前人们就已经知道，不同作物种类在其可食用部分积累镉的能力有所不同。例如，绿叶蔬菜通常比谷物或水果作物积累更高浓度的镉（Bingham 等，1975；Chaney 和 Hornick，1977）。如果土壤被镉污染，农民可以选择改变在特定土地上种植的作物类型。如果没有选择性，可以在同一作物类型下选择对镉具有低积累性的作物品种。众所周知，同一类植物的不同品种积累镉的速率不同，这可能与生根方式不同、根系吸收镉的方式不同或植物体内镉转运方式不同有关（Grant 等，2008）。在一些国家，专门培育的镉低吸收性的品种也随之商业化（Clarke 等，1997），而在其他地区，农民可以从可获得的商业品种中选择吸收镉较低的品种（这些信息是可以获得的）。通过选择适当的作物轮作计划，也可以尽量减少镉对食物链的污染；有证据表明，某些作物的轮作顺序（如在羽扇豆种植后种植小麦）可能会促进镉的积累（Oliver 等，1993），目前原因尚不清楚，这可能与土壤化学或物理条件的改变（如土壤pH的变化）有关。最后，农民也可以选择种植可以从土壤中提取镉的植物（即植物提取方法），并在种植粮食作物之前处理掉这些植物（Murakami 等，2009）。这一战略现在已经成熟且到了具有实际可行性的阶段（Abe 等，2017）。

与此同时，也可通过选择或控制土壤的化学和物理条件来尽量减少镉在食物链中的积累。不同的场地可能具有不同的土壤条件：具有高pH、高黏土、高有机质、高锌和低镉含量的土壤更有可能在作物中产生最小量的镉积累（Grant 等，1999）。如果无法选择场地，则可以尝试土壤处理。由于镉是一种阳离子金属，通过添加石灰提高土壤pH以及增加土壤阳离子交换能力，可以增加土壤吸附并减少作物吸收，但是以上方法在田间研究中效果并不一致。与此同时，也可以在土壤中添加吸附剂，以增强对镉的结合，将作物对镉的吸收减少到最低限度（Komárek, Vaněk 和 Ettler, 2013；Tang 等，2016），但是这通常需要较高的施用量（每公顷数吨）且修复的时效是未知的。研究还表明，添加锌也可以降低作物中镉的浓度（Oliver 等，1994），主要是由于锌对可食用部分中的镉存在竞争性吸收（Welch 等，1999）。最后，如果镉污染是人为的而不是自然原因造成的，很可能污染仅限于表层土壤。对于许多污染物，耕作、埋藏或稀释污染层会减少作物对镉的吸收，因为大多数作物的根只活跃在土壤上层的10~20cm。

避免使用富含氯的灌溉水也会减少镉对食物链的污染，因为氯可以结合Cd^{2+}离子，增加了其在土壤中的流动性，从而增加了植物对镉的吸收（McLaughlin 等，1994）。

3.3.4 类金属

砷（As）是农业土壤中分布最广、危害最严重的类金属污染物，其自然源比人为源更为广泛（Bhattacharya 等，2007）。砷对食物链的污染主要发生在水稻种植系统中，在水稻土壤中，低氧化还原条件通过溶解与砷结合的氧化铁矿物来使砷具有流动性，同时还能将砷酸盐离子还原成亚砷酸盐，亚砷酸盐在土壤中比砷酸盐流动性更强（Hamon 等，2004）。由于土壤中的这些化学反应以及根系吸收途径的存在，可以通过严格的水管理（升高床体、季中排水或旱地耕作）增加土壤氧化还原性（Hu 等，2013），以及通过硅肥的添加来使水稻中砷的积累最小化。然而，相比于水稻淹水栽培，好氧水稻栽培的缺点是镉积累可能会增加（Hu 等，2013）。选择合适的品种可以减少稻米中的砷（Norton 等，2009）。

3.3.5 放射性核素

减少放射性核素在食物链中积累的农业操作措施主要源自对切尔诺贝利、戈亚尼亚和福岛核事故影响的研究（Fesenko 等，2017）。受关注的主要同位素是污染事件发生后早期的^{131}I，以及后期的铯和锶同位素（^{134}Cs、^{137}Cs 和 ^{90}Sr）。^{131}I半衰期非常短，为8.02天，其主要风险途径是牧草-奶牛-牛奶-人类链。因此，在发生含^{131}I污染事件后，需要立即采取的主要管理措施是限制动物进入受污染的牧场，如果可能的话需要从污染区域以外运送饲料进行喂养。对于呈阳离子态的铯和锶的放射性同位素，其修复措施与镉相似，可以采取更换作物种类和栽培品种、采用高CEC的吸附剂、施用石灰和施肥管理等措施（Fesenko 等，2007）。对于Cs，施用钾肥对减少植株吸收Cs特别有效，这是由于钾离子与Cs离子在根系吸收方面存在竞争关系（Shaw, 1993），而钙对减少^{90}Sr的吸收特别有效（Nisbet 等，1993）。但是应避免使用氨肥，因为它可能会增加对^{137}Cs和^{90}Sr的吸收（Guillén 等，2017）。土壤翻耕、土壤移除、将表层污染埋藏到更深的土层中也可稀释或降低土壤中的同位素浓度（Fesenko 等，2017）。

4 土壤污染与修复案例研究

4.1 联合国外地特派团通过加强自然衰减来进行POL污染地的修复：关于联合国在科特迪瓦行动的个案研究[18]

在外场任务中使用石油和润滑油（POL）是不可避免的，因为需要利用它们进行发电和操作机械设备以保障维和行动。这些可能对环境造成重大影响的过程，增加了土壤污染的可能性。本节介绍的是全球服务中心/环境技术支援小组在联合国结束一项在科特迪瓦的外地行动后，对POL污染场地进行修复工作的个案研究。

该项目的目标是将受污染土壤的总石油烃含量从36 000毫克/千克至75 000毫克/千克降低到400毫克/千克至1 000毫克/千克的背景总石油烃含量水平，为植被恢复提供有利的环境。在这项工程中，超过1 200吨被POL污染的土壤从场地被移走，并用新鲜的土壤替代，挖掘出的污染土壤使用当地的天然材料进行处理。

被污染的土壤存放在一个大型混凝土搅拌机中进行搅拌和曝气，以促进微生物生长和POL的分解。两种成分[家禽粪便和天然表面活性物质或棕榈灰皂（也称为黑皂）]被添加到混合物中，以改善土壤条件并加速微生物修复。

结果表明，修复后的总石油烃水平初期就降低95%以上，而在天然微生物的活性得以保持的情况下，14天内总石油烃水平得到持续降低。随后在被恢复的地区种植了当地草本。这个案例研究强调了以低成本的修复技术在联合国外地任务中减轻POL污染的重要性。

---

18 环境技术支援组（ETSU）（GSC环境技术支援组，意大利阿普利亚）。

土壤污染：一个隐藏的现实

4.2 西伯利亚西部针叶林带利用当代修复方法进行石油污染土地的修复[19]

俄罗斯在全球石油生产中占有领先地位，其70%以上的石油是在西伯利亚西部的针叶林地区开采出来的。20世纪90年代，该地区石油生产企业发生的管道破裂事故频率急剧增加，造成了生态系统的石油污染。在国家对环境保护立法的法定合规性控制不足情况下，大量受石油污染的土地长时间没有得到修复，这成为目前在该地区从事石油生产的新公司的"历史遗留"问题。

在过去的10～15年里，石油公司为恢复受石油污染的土地做出了巨大的努力，但这个问题并没有完全被解决。主要是由于该地区特殊的环境条件：其年平均气温在-0.1℃～5℃，1月份平均气温为-18℃～-24℃（最低记录为-62℃）；积雪稳定覆盖期可达180～200天；降水量大大超过了蒸发量。西伯利亚西部低地是一片广阔的支离破碎的平原，在全新世期间经历了沼泽形成的快速发展，在一些地区，沼泽覆盖了90%的领土。石油泄漏主要发生在湿地生态系统中，使得相应的机械修复作业非常复杂化。

在湿地生态系统中，不仅有不利的天气条件，而且采用的修复技术也不适用于湿地土壤，因为这些技术最初是为矿质土壤开发的。基本的技术解决方案包括表面石油收集（如果有的话）、农业技术措施（石灰、矿物施肥）、生物刺激（天然石油氧化微生物的活化）或生物强化（应用具有石油氧化作用的商业生物产品）、周期性松土和植物改良（草甸牧草播种）。然而，对于石油污染泥炭沼泽土壤的修复，还需要一些其他的方法。

泥炭土对石油有很强的吸附能力，因此在泄漏后立即收集溢出的石油是非常困难的，在石油变稠之后也是不可能的。与此同时，石油在泥炭沼泽土壤剖面的污染集中于土壤上层，其石油碳氢化合物的浓度可达80%或更多，这远远超过了微生物分解所能消耗的水平。即使经过几个季节的反复处理，上述传统技术解决方案仍缺乏显著效果。

如果在石油污染区域，使用机械首先清除最上层的污染层（通常10～15厘米），可显著提高回收效率。在这一层中，除了重油碳氢化合物外，还积累了大量的树脂和沥青质。这种积累有效地密封了土壤，阻止了水和气体的转移交换。这反过来极大地降低了污染土壤中的微生物活性。在实施这项技术操作的最初阶段，通常要投入大量的人工劳动。这就解释了为什么在某些石油污染地点中修复效率很高，但再生土壤的总面积仍然很低。

---

19 Sergey Trofimov, Ruslan Kinjaev, Olga Yakimenko （莫斯科国立罗蒙诺索夫大学土壤系，俄罗斯，莫斯科）

后来，人们开始使用挖掘机进行相关的工作（图21），这使得每年被石油污染的土地的再生面积成倍增加成为可能。

在上层土壤被清除后，土壤中石油碳氢化合物的浓度通常不会超过微生物分解石油的能力水平；这样就可以允许使用传统的生物复垦方法。然而，将石油碳氢化合物浓度进一步降低到可接受的水平仍然是一项艰巨的任务。

对于石油污染土壤修复，最重要的问题之一是土壤酸碱性的调控。细菌分解石油的最佳pH通常为6～8，而泥炭土的pH一般在3.5～4.5，具有较高的交换性和酸性。因此，为了达到最佳pH所必须添加的大量石灰，使得这项任务在技术上和经济上都不可行且不合理。

图21 浮动式挖掘机作业（Pxhere）

解决这一问题的方法之一是使用生物降解剂，它能够在pH 4～4.5时氧化碳氢化合物。然而，为了有效地降解石油，必须为沼泽泥炭土壤提供适当的曝气，这在实践中是极其难实现的。为了克服这个问题，结合使用生物强化和植物改良技术似乎有可能解决此问题（Glick，2003；Khan等，2013）。这种结合将在生物降解剂中的微生物和沼泽植物之间提供一种共生互动，从而将空气通过薄壁组织传输到根系，随后空气中的氧气扩散进入到根际，这为分解细菌氧化石油提供了可能性。

除了提供氧气，植物还能通过根系分泌物刺激根际微生物群的功能（Bais等，2006）。反过来，细菌可以通过产生各种植物激素和抗胁迫物质来刺激植物的发育（Safronova等，2006），从而使植物即使在石油重污染的条件下也能生长。此外，细菌可以固定分子氮，使水解磷酸盐流动，产生铁载体，这些也可以促进植物发育。然而，目前还没有生产出能同时具有上述所有功能的生物降解剂。这使得开发和在实际生产中运用合适的生物降解剂变得极其紧迫，正如开展适合西伯利亚西部针叶林地区典型沼泽的植物育种一样紧迫。

4.3 辅助植物固定：西班牙东南部尾矿的一种有效修复技术[20,21]

采矿在卡塔赫纳-拉尤尼翁山脉（穆尔西亚，西班牙）已经存在了2500多年，在这一地带，由于硫化锌和硫化铅等硫化物矿物的开采产生了大量尾矿。尾矿池在1991年活动停止后被废弃，其由于涉及有毒金属（类金属）含量高的风险，引起了人们的极大关注。

此外，这些尾矿肥力低，有机质含量低，而且酸性高。因此，除非添加有机或无机改性剂，否则要建立原生植被是非常困难的（García 和 Lobo，2013）。植物修复被认为是修复土壤污染的一种经济、环保的方法（Wan, Lei 和 Chen, 2016）。而在植物修复技术中，辅助植物固定是降低污染物扩散风险的一种解决方法（Yang 等，2016）。相关研究提出了几种类型的土壤改良剂来稳定土壤中的金属（类金属），并提出了若干修正意见（Kumpiene, Lagerkvist 和 Maurice, 2008）。有机改良剂和富含碳酸盐的材料已被成功地用于降低金属的生物有效性和恢复受污染土壤的生态功能中（Park 等，2011）。

本研究的主要目的是确定在铅锌矿尾库复垦30个月后对其进行辅助植物固定处理的有效性。通过监测尾矿的理化生化特性和生物可利用金属（类金属）（As、Cd、Pb和Zn）含量，对该方法的有效性进行了评价。此外，研究者也对金属（类金属）向植物根、茎、叶的转移和植物群落的进化进行了评估。最初的假设是，在无机和有机改良剂施用条件下，利用原生植物实施植物固定，会降低金属的流动性，降低环境和公共健康风险，并提高土壤质量、肥力和植被覆盖。植物会在根部积累高浓度的金属（类金属），而向冠部的迁移率比较低。

这项研究是在位于卡塔赫纳-拉尤尼翁矿区的Santa Antonieta尾矿池进行的。这个尾矿池占地有1.4公顷。大理石废料被用作碳酸盐的来源，以中和酸性、固定金属和改善土壤结构。猪粪浆及其固相（粪便）经物理分离后作为土壤开发和植被建立的有机质和养分来源。

以下植物种类在2012年被种植于该地：滨藜、岩蔷薇、蜡菊、海雀花、薰衣草、利坚草、迷迭香、棉毛菊属、落芒草属、百慕大草、补血草、苦苣菜。

20　S. Martínez-Martínez, R. Zornoza, J.A. Acosta, M. Gabarrón, M.D.Gómez-López 和 A. Faz（西班牙卡塔赫纳理工大学水土资源可持续利用、管理与复垦研究小组）

21　致谢:本项目由欧盟LIFE+项目资助（LIFE09 ENV/ES/000439）。

研究结果表明，结合大理石废渣、猪粪浆和粪肥施用，利用辅助植物来恢复酸性尾矿池是有效的。该技术提高了土壤pH、阳离子交换能力、土壤总有机碳和养分含量，改善了土壤结构，可使镉、铅、锌等金属的流动性降低90%～99%。利坚草和落芒草属对铅、锌和砷的植物稳定作用显著，其根系金属含量较高，很少向冠层分配，而滨藜和棉毛菊属叶子中的锌具有植物毒性。因此，应避免在高锌污染的土壤中种植这些品种。

参考文献

Aarts, H. F. M., Habekotté, B. &Keulen, H. van. 2000. Phosphorus(P)managementinthe 'De Marke' dairy farming system. *Nutrient Cycling in Agroecosystems*, 56(3): 219–229. https: //doi. org/10. 1023/A: 1009814905339.

Abdel-Shafy, H. I. &Mansour, M. S. M. 2016. Areviewonpolycyclicaromatichydrocarbons: Source, environmental impact, effect on human health and remediation. *Egyptian Journal of Petroleum*, 25(1): 107–123. https: //doi. org/10. 1016/j. ejpe. 2015. 03. 011.

Abe, T., Ito, M., Takahashi, R., Honma, T., Sekiya, N., Shirao, K., Kuramata, M., Murakami, M. &Ishikawa, S. 2017. Breedingofapracticalriceline 'TJTT8' forphytoextraction of cadmium contamination in paddy fields. *Soil Science* and *Plant Nutrition*, 63(4): 388–395. https: //doi. org/10. 1080/00380768. 2017. 1345598.

Abrahams, P. W. 2002. Soils: theirimplicationstohumanhealth. *Science of The Total Environment*, 291(1–3): 1–32. https: //doi. org/10. 1016/S0048-9697(01)01102-0.

Absalom, J. P., Young, S. D. &Crout, N. M. J. 1995. Radio-caesiumfixationdynamics: measurement in six Cumbrian soils. *European Journal of Soil Science*, 46(3): 461–469. https: //doi. org/10. 1111/j. 1365-2389. 1995. tb01342. x.

Absalom, J. P., Young, S. D., Crout, N. M. J., Nisbet, A. F., Woodman, R. F. M., Smolders, E.&Gillett, A. G. 1999. Predicting Soil to Plant Transfer of Radiocesium Using Soil Characteristics. *Environmental Science &Technology*, 33(8): 1218–1223. https: //doi. org/10. 1021/es9808853.

Ahmad, M. S. &Ashraf, M. 2011. Essentialrolesandhazardouseffectsofnickelin plants. *Reviews of environmental contamination* and *toxicology*, 214: 125–167. https: //doi. org/10. 1007/978-1-4614-0668-6_6.

Aichner, B., Bussian, B., Lehnik-Habrink, P. &Hein, S. 2013. Levelsand Spatial Distribution of Persistent Organic Pollutants in the Environment: A Case Study of German Forest Soils. *Environmental Science &Technology*, 47(22): 12703–12714. https: //doi. org/10. 1021/es4019833.

Aktar, W., Sengupta, D. &Chowdhury, A. 2009. Impactofpesticidesuseinagriculture: their benefits and hazards. *Interdisciplinary Toxicology*, 2(1): 1–12. https: //doi. org/10. 2478/v10102-009-0001-7.

Albanese, S., DeVivo, B., Lima, A. &Cicchella, D. 2007. Geochemicalbackground and baseline values of toxic elements in stream sediments of Campania region (Italy). *Journal of Geochemical Exploration*, 93(1): 21–34. https: //doi. org/10. 1016/j. gexplo. 2006. 07. 006.

Albert, R. E. 1987. Issuesinbiochemicalapplicationstoriskassessment: howdowe predicttoxicityofc omplexmixtures?*Environmental Health Perspectives*, 76: 185–186.

Alekshun, M. N. &Levy, S. B. 2007. MolecularMechanismsofAntibacterialMultidrug Resistance. *Cell*, 128(6): 1037–1050. https: //doi. org/10. 1016/j. cell. 2007. 03. 004.

Alexander, M. 1999. *Biodegradation* and *bioremediation*. 2nd ed edition. SanDiego, Academic Press. 453pp.

Allen, H. K., Donato, J., Wang, H. H., Cloud-Hansen, K. A., Davies, J. &Handelsman, J. 2010. Call of the wild: antibiotic resistance genes in natural environments. *Nature Reviews Microbiology*, 8(4): 251–259. https: //doi. org/10. 1038/nrmicro2312.

Allende, A. &Monaghan, J. 2015. IrrigationWaterQualityforLeafyCrops: APerspective of Risks and Potential Solutions. *International Journal of Environmental Research* and *PublicHealth*, 12(7): 7457–7477. https: //doi. org/10. 3390/ijerph120707457.

Alloway, B. J., ed. 2013. *HeavyMetalsinSoils: TraceMetalsandMetalloidsinSoilsand their Bioavailability*. Third edition. Environmental Pollution. Springer Netherlands. (alsoavailableat// www. springer. com/gp/book/9789400744691).

Amanullah. 2015. TheRoleofBeneficialMicrobes(Biofertilizers)InIncreasingCrop Productivity and Profitability. *EC Agriculture 2. 6*: 504.

Amanullah &Inamullah. 2016. Residual phosphorus and zinc influence wheat productivity under rice–wheat cropping system. *SpringerPlus*, 5(1). https: //doi. org/10. 1186/s40064-016-1907-0.

Amanullah, J. & Stewart, B. A. 2013. Dry Matter Partitioning, Growth Analysis and WaterUs eEfficiencyResponseofOats(AvenasativaL.)toExcessiveNitrogenand Phosphorus Application. *Journal of Agricultural Science* and *Technology*, 15(3): 479–489.

Amanullah, Khan, S. -T., Iqbal, A. &Fahad, S. 2016. GrowthandProductivityResponseof Hybrid Rice to Application of Animal Manures, Plant Residues and Phosphorus. *Frontiers in Plant Science*, 7. https: //doi. org/10. 3389/fpls. 2016. 01440.

AMAP. 1997. ArcticPollutionIssues: AStateoftheArcticEnvironmentReport. Oslo, ArticMonitor ingandAssessmentProgramme. (alsoavailableathttps: //www. amap. no/documents/doc/arctic-pollution-issues-a-state-of-the-arctic-environment- report/67).

Anda, M. 2012. Cationimbalanceandheavymetalcontentofsevenindonesiansoils as affected by elemental compositions of parent rocks. *Geoderma*, 189–190: 388–396. https: //doi. org/10. 1016/j. geoderma. 2012. 05. 009.

Andersson, J. T. &Achten, C. 2015. TimetoSayGoodbyetothe16EPAPAHs?Toward an Up-to-Date Use of PACs for Environmental Purposes. *Polycyclic Aromatic Compounds*, 35(2–4): 330–354.

https: //doi. org/10. 1080/10406638. 2014. 991042.

Appel, C. &Ma, L. 2002. Concentration, pH, andSurfaceChargeEffectsonCadmium andLeadSorpt ioninThreeTropicalSoils. *J. ENVIRON. QUAL.*, 31: 9.

Araújo, P. H. H., Sayer, C., Giudici, R. &Poço, J. G. R. 2002. Techniquesforreducingresidual monomer content in polymers: A review: Techniques for Reducing Residual Monomer Content. *Polymer Engineering & Science*, 42(7): 1442–1468. https: //doi. org/10. 1002/pen. 11043.

Arias-Estévez, M., López-Periago, E., Martínez-Carballo, E., Simal-Gándara, J., Mejuto, J. C. &García-Río, L. 2008. Themobilityanddegradationofpesticidesinsoilsandthe pollution of groundwater resources. *Agriculture, Ecosystems &Environment*, 123(4): 247–260. https: //doi. org/10. 1016/j. agee. 2007. 07. 011.

Arikan, O. A., Mulbry, W. &Rice, C. 2009. Managementofantibioticresiduesfrom agricultural sources: Use of composting to reduce chlortetracycline residues in beef manure from treated animals. *Journal of Hazardous Materials*, 164(2–3): 483–489. https: //doi. org/10. 1016/j. jhazmat. 2008. 08. 019.

AustralianGovernment. 2018. *Phthalates—finalhazardassessmentandcompendium (NICNAS)* [online]. [Cited 3 April 2018]. https: //www. nicnas. gov. au/chemical-information/other-assessments/reports/phthalates-hazard-assessments.

Azanu, D., Mortey, C., Darko, G., Weisser, J. J., Styrishave, B. &Abaidoo, R. C. 2016. Uptake of antibiotics from irrigation water by plants. *Chemosphere*, 157: 107–114. https: //doi. org/10. 1016/j. chemosphere. 2016. 05. 035.

Baderna, D., Maggioni, S., Boriani, E., Gemma, S., Molteni, M., Lombardo, A., Colombo, A., Bordonali, S., Rotella, G., Lodi, M. &Benfenati, E. 2011. A combined approach to investigate the toxicity of an industrial landfill's leachate: Chemical analyses, risk assessment and in vitro assays. *Environmental Research*, 111(4): 603–613. https: //doi. org/10. 1016/j. envres. 2011. 01. 015.

Bais, H. P., Weir, T. L., Perry, L. G., Gilroy, S. &Vivanco, J. M. 2006. THEROLEOFROOT EXUDATES IN RHIZOSPHERE INTERACTIONS WITH PLANTS ANDOTHER ORGANISMS. *Annual Review of Plant Biology*, 57(1): 233–266. https: //doi. org/10. 1146/ annurev. arplant. 57. 032905. 105159.

Baldantoni, D., Morra, L., Zaccardelli, M. &Alfani, A. 2016. Cadmiumaccumulationin leaves of leafy vegetables. *Ecotoxicology* and *Environmental Safety*, 123: 89–94. https: // doi. org/10. 1016/j. ecoenv. 2015. 05. 019.

Bányiová, K., erná, M., Mikeš, O., Komprdová, K., Sharma, A., Gyalpo, T., upr, P. &Scheringer, M. 2017. Long-term time trends in human intake of POPs in the Czech Republic indicate a need for continuous monitoring. *Environment International*, 108: 1–10. https: //doi.

org/10. 1016/j. envint. 2017. 07. 008.

Barba-Gutiérrez, Y., Adenso-Díaz, B. &Hopp, M. 2008. An analysis ofsomeenvironmental consequences of European electrical and electronic waste regulation. *Resources, Conservation&Recycling*, 3(52): 481–495. https: //doi. org/10. 1016/j. resconrec. 2007. 06. 002.

Basile, B. P., Middleditch, B. S. &Oró, J. 1984. Polycyclicaromatichydrocarbonsinthe Murchisonmeteorite. *OrganicGeochemistry*, 5(4): 211–216. https: //doi. org/10. 1016/0146-6380(84)90008-1.

Bauman-Kaszubska, H. & Sikorski, M. 2009. Selected problems of waste water disposal and sludge handling in the Mazovian province. *Journal of Water* and *Land Development*, 13b(1). https: //doi. org/10. 2478/v10025-010-0011-z.

Bayat, J., Hashemi, S. H., Khoshbakht, K. &Deihimfard, R. 2016. Fingerprinting aliphatic hydrocarbon pollutants over agricultural lands surrounding Tehran oil refinery. *Environmental Monitoring* and *Assessment*, 188(11). https: //doi. org/10. 1007/s10661-016- 5614-7.

Bell, J. N. B., Minski, M. J. &Grogan, H. A. 1988. Plantuptakeofradionuclides. *SoilUseand Management*, 4(3): 76–84. https: //doi. org/10. 1111/j. 1475-2743. 1988. tb00740. x.

Berends, G. &Kobayashi, M. 2012. FoodafterFukushima-Japan'sRegulatoryResponse to the Radioactive Contamination of Its Food Chain. *Food* and *Drug Law Journal*, 67: 51.

Bernhardt, A. &Gysi, N. 2016. World'sWorstPollutionProblems. Thetoxicbeneath our feet., p. 56. Green Cross Switzerland and Pure Earth Foundation. (also available at http: //www. worstpolluted. org/docs/WorldsWorst2016. pdf).

Beuchat, L. R. 2002. Ecologicalfactorsinfluencingsurvivalandgrowthofhuman pathogens on raw fruits and vegetables. *Microbes* and *Infection*, 4(4): 413–423.

Beyer, W. N. 1990. EvaluatingSoilContamination., p. 25. No. 90(2). USFishWildlife Service. (also available at https: //www. nwrc. usgs. gov/wdb/pub/others/FWS_Bio_ Rep_90-2. pdf).

BeyondPesticides. 2018. WhatIsIntegratedPestManagement? In: *BeyondPesticides* [online]. [Cited 3 April 2018]. https: //beyondpesticides. org/resources/safety-source- on-pesticide-providers/what-is-integrated-pest-management.

Bhatia, R., Shiau, R., Petreas, M., Weintraub, J. M., Farhang, L. &Eskenazi, B. 2005. Organochlorine Pesticides and Male Genital Anomalies in the Child Health and Development Studies. *Environmental Health Perspectives*, 113(2): 220–224. https: //doi. org/10. 1289/ehp. 7382.

Bhattacharya, P., Welch, A. H., Stollenwerk, K. G., McLaughlin, M. J., Bundschuh, J. &Panaullah, G. 2007. Arsenicintheenvironment: BiologyandChemistry. *TheScienceoftheTotal Environment*, 379(2–3): 109–120. https: //doi. org/10. 1016/j. scitotenv. 2007. 02. 037.

Bien, J., Neczaj, E. &Milczarek, M. 2013. CO – COMPOSTING OF MEAT PACKING

WASTEWATER SLUDGE and ORGANIC FRACTION OF MUNICIPAL SOLID WASTE. *GlobalNESTJournal*, 15(4): 513–521.

Bingham, F. T., Page, A. L., Mahler, R. J. &Ganje, T. J. 1975. Growthand Cadmium Accumulation of Plants Grown on a Soil Treated with a Cadmium-Enriched Sewage Sludge 1. *Journal of Environmental Quality*, 4(2): 207–211. https: //doi. org/10. 2134/ jeq1975. 00472425000400020015x.

Björnsdotter, M. 2015. *Leachingofresidualmonomers, oligomersandadditivesfrom polyethylene, polypropylene, polyvinyl chloride, high-density polyethylene* and *polystyrene virgin plastics.* ÖrebroUniversity. (also available at https: //www. diva- portal. org/smash/get/diva2: 855478/ FULLTEXT01. pdf).

Blaser, P., Zimmermann, S., Luster, J. &Shotyk, W. 2000. Critical examination of trace element enrichments and depletions in soils: As, Cr, Cu, Ni, Pb, and Zn in Swissforestsoils. *TheScienceof thetotalenvironment*, 249(1–3): 257–280. https: //doi. org/10. 1016/S0048-9697(99)00522-7.

Blum, W. E. H. 2005. FunctionsofSoilforSocietyandtheEnvironment. *Reviewsin EnvironmentalScienceandBio/Technology*, 4(3): 75–79. https: //doi. org/10. 1007/s11157- 005- 2236-x.

Blume, H. -P., Brümmer, G. W., Fleige, H., Horn, R., Kandeler, E., Kögel-Knabner, I., Kretzschmar, R., Stahr, K. &Wilke, B. -M. 2016. *Scheffer/SchachtschabelSoilScience*. BerlinHeidelberg, Springer-Verlag. (alsoavailableat//www. springer. com/us/ book/9783642309410).

deBoer, J. &Fiedler, H. 2013. Persistentorganicpollutants. *TrACTrendsinAnalytical Chemistry*, 46: 70–71. https: //doi. org/10. 1016/j. trac. 2013. 03. 001.

Bolan, N., Kunhikrishnan, A., Thangarajan, R., Kumpiene, J., Park, J., Makino, T., Kirkham, M. B. &Scheckel, K. 2014. Remediationofheavymetal(loid)scontaminatedsoils–To mobilize or to immobilize? *Journal of Hazardous Materials*, 266: 141–166. https: //doi. org/10. 1016/j. jhazmat. 2013. 12. 018.

Bolívar, J. P., García-Tenorio, R. &García-León, M. 1995. Enhancement ofnaturalradioactivity in soils and salt-marshes surrounding a non-nuclear industrial complex. *Science of The Total Environment*, 173–174: 125–136. https: //doi. org/10. 1016/0048-9697(95)04735-2.

Bondarczuk, K., Markowicz, A. &Piotrowska-Seget, Z. 2016. The urgent need forriskassessment on the antibiotic resistance spread via sewage sludge land application. *Environment International*, 87: 49–55. https: //doi. org/10. 1016/j. envint. 2015. 11. 011.

Bossi, R., Dam, M. &Rigét, F. F. 2015. Perfluorinatedalkylsubstances(PFAS)interrestrial environments in Greenland and Faroe Islands. *Chemosphere*, 129: 164–169. https: // doi. org/10. 1016/j. chemosphere. 2014. 11. 044.

Boxall, A. B. A., Johnson, P., Smith, E. J., Sinclair, C. J., Stutt, E. &Levy, L. S. 2006. Uptake of Veterinary Medicines from Soils into Plants. *Journal of Agricultural* and *Food Chemistry*, 54(6): 2288–2297. https: //doi. org/10. 1021/jf053041t.

Boxall, A. B. A., Rudd, M. A., Brooks, B. W., Caldwell, D. J., Choi, K., Hickmann, S., Innes, E., Ostapyk, K., Staveley, J. P., Verslycke, T., Ankley, G. T., Beazley, K. F., Belanger, S. E., Berninger, J. P., Carriquiriborde, P., Coors, A., DeLeo, P. C., Dyer, S. D., Ericson, J. F., Gagné, F., Giesy, J. P., Gouin, T., Hallstrom, L., Karlsson, M. V., Larsson, D. G. J., Lazorchak, J. M., Mastrocco, F., McLaughlin, A., McMaster, M. E., Meyerhoff, R. D., Moore, R., Parrott, J. L., Snape, J. R., Murray- Smith, R., Servos, M. R., Sibley, P. K., Straub, J. O., Szabo, N. D., Topp, E., Tetreault, G. R., Trudeau, V. L. &VanDerKraak, G. 2012. PharmaceuticalsandPersonalCareProductsin the Environment: What Are the Big Questions?*Environmental Health Perspectives*, 120(9): 1221–1229. https: //doi. org/10. 1289/ehp. 1104477.

Bradl, H. B. 2004. Adsorptionofheavymetalionsonsoilsandsoilsconstituents. *Journal ofColloidandInterfaceScience*, 277(1): 1–18. https: //doi. org/10. 1016/j. jcis. 2004. 04. 005.

Bragazza, L., Freeman, C., Jones, T., Rydin, H., Limpens, J., Fenner, N., Ellis, T., Gerdol, R., Hajek, M., Hajek, T., Iacumin, P., Kutnar, L., Tahvanainen, T. &Toberman, H. 2006. Atmospheric nitrogen deposition promotes carbon loss from peat bogs. *Proceedings of the National AcademyofSciences*, 103(51): 19386–19389. https: //doi. org/10. 1073/pnas. 0606629104.

Brahushi, F., Dörfler, U., Schroll, R. & Munch, J. C. 2004. Stimulation of reductive dechlorinationofhexachlorobenzeneinsoilbyinducingthenativemicrobialactivity. *Chemosphere*, 55(11): 1477–1484. https: //doi. org/10. 1016/j. chemosphere. 2004. 01. 022.

Brand, E., Otte, P. &Lijzen, J. 2007. CSOIL2000anexposuremodelforhumanrisk assessment of soil contamination. A model description. (also available at http: // rivm. openrepository. com/ rivm/handle/10029/258236).

Brevik, E. C. 2013. Soilsandhumanhealth: Anoverview. *In*E. C. Brevik&L. C. Burgess, eds. *Soils and human health*, pp. 29–58.

Brinkac, L., Voorhies, A., Gomez, A. &Nelson, K. E. 2017. TheThreatofAntimicrobial Resistance on the Human Microbiome. *Microbial Ecology*, 74(4): 1001–1008. https: // doi. org/10. 1007/s00248-017-0985-z.

Brodesser, J., Byron, D. H., Cannavan, A., Ferris, I. G., Gross-Helmet, K., Hendrichs, J., Maestroni, B. M., Unsworth, J., Vaagt, G. & Zapata, F. 2006. Pesticides in developing countries and the International Code of Conduct on the Distribution and the Use of Pesticides. *Food* and *Environmental Protection Newsletter*, 9(2). (also available at http: //www-naweb. iaea. org/nafa/fep/public/fep-nl-9-2. pdf).

Brody, J. G., Moysich, K. B., Humblet, O., Attfield, K. R., Beehler, G. P. &Rudel, R. A. 2007. Environmental pollutants and breast cancer: Epidemiologic studies. *Cancer*, 109(S12): 2667–2711. https: //doi. org/10. 1002/cncr. 22655.

Browne, M. A., Dissanayake, A., Galloway, T. S., Lowe, D. M. &Thompson, R. C. 2008. Ingested microscopic plastic translocates to the circulatory system of the mussel, Mytilus edulis(L). *EnvironmentalScience&Technology*, 42(13): 5026–5031.

Brzóska, M. M. &Moniuszko-Jakoniuk, J. 2005. Disordersinbonemetabolismoffemale rats chronically exposed to cadmium. *Toxicology* and *Applied Pharmacology*, 202(1): 68–83. https: //doi. org/10. 1016/j. taap. 2004. 06. 007.

Bundschuh, J., Litter, M. I., Parvez, F., Román-Ross, G., Nicolli, H. B., Jean, J. -S., Liu, C. -W., López, D., Armienta, M. A., Guilherme, L. R. G., Cuevas, A. G., Cornejo, L., Cumbal, L. &Toujaguez, R. 2012. OnecenturyofarsenicexposureinLatinAmerica: areviewofhistoryand occurrencefrom14countries. *TheScienceoftheTotalEnvironment*, 429: 2–35. https: // doi. org/10. 1016/j. scitotenv. 2011. 06. 024.

Bünemann, E. K., Schwenke, G. D. &VanZwieten, L. 2006. Impactofagriculturalinputs on soil organisms—a review. *Australian Journal of Soil Research*, 44(4): 379. https: //doi. org/10. 1071/SR05125.

Burgess, L. C. 2013. Organicpollutantsinsoil. *Soilsandhumanhealth*, pp. 83–106. Boca Raton, Fla, CRCPress.

Buser, H. -R., Poiger, T. &Müller, M. D. 1999. OccurrenceandEnvironmentalBehaviorof the Chiral Pharmaceutical Drug Ibuprofen in Surface Waters and in Wastewater. *Environmental Science &Technology*, 33(15): 2529–2535. https: //doi. org/10. 1021/ es981014w.

ButlerEllis, M. C., vandeZande, J. C., vandenBerg, F., Kennedy, M. C., O'Sullivan, C. M., Jacobs, C. M., Fragkoulis, G., Spanoghe, P., Gerritsen-Ebben, R., Frewer, L. J. &Charistou, A. 2017. The BROWSE model for predicting exposures of residents and bystanders to agricultural use of plant protection products: An overview. *Biosystems Engineering*, 154: 92–104. https: //doi. org/10. 1016/j. biosystemseng. 2016. 08. 017.

Cachada, A., FerreiradaSilva, E., Duarte, A. C. &Pereira, R. 2016. Riskassessmentofurban soils contamination: The particular case of polycyclic aromatic hydrocarbons. *Science of The Total Environment*, 551–552: 271–284. https: //doi. org/10. 1016/j. scitotenv. 2016. 02. 012.

Cachada, A., Rocha-Santos, T. &Duarte, A. C. 2018. Chapterl-SoilandPollution: An IntroductiontotheMainIssues. *SoilPollution*, pp. 1–28. AcademicPress. (alsoavailable athttps: // www. sciencedirect. com/science/article/pii/B9780128498736000017).

Cai, D. W. 2008. Understandtheroleofchemicalpesticidesandpreventmisusesof pesticides. *BulletinofAgriculturalScienceandTechnology*, 1: 36–38.

Callahan, M. A. &Sexton, K. 2007. IfCumulativeRiskAssessmentIstheAnswer, What Is the Question? *Environmental Health Perspectives*, 115(5): 799–806. https: //doi. org/10. 1289/ehp. 9330.

Cameron, K. C., Di, H. J. &Moir, J. L. 2013. Nitrogenlossesfromthesoil/plantsystem: a review: Nitrogen losses. *Annals of Applied Biology*, 162(2): 145–173. https: //doi. org/10. 1111/aab. 12014.

Canadian Council of Ministers of the Environment. 1999. Guidance Manual for Developing Site-Specific Soil Quality Remediation Objectives for Contaminated Sites in Canada. http: // ceqg-rcqe. ccme. ca/download/en/251?redir=1522869161.

Cang, L., Wang, Y., Zhou, D. &Dong, Y. 2004. Heavymetalspollutioninpoultryand livestock feeds and manures under intensive farming in Jiangsu Province, China. *JournalofEnvironmental Sciences(China)*, 16(3): 371–374.

Carlon, C., European Commission, Joint Research Centre & Institute for Environment and Sustainability. 2007. *Derivation methods of soil screening values in Europe: a review of national procedures towards harmonisation : a report of the ENSURE Action.* Luxembourg, EUR-OP.

Carpenter, S. R., Caraco, N. F., Correll, D. L., Howarth, R. W., Sharpley, A. N. &Smith, V. H. 1998. NONPOINT POLLUTION OF SURFACE WATERSWITHPHOSPHORUS ANDNITROGEN. *EcologicalApplications*, 8(3): 559–568. https: //doi. org/10. 1890/1051-0761(1998)008[0559: NPOSWW]2. 0. CO; 2.

Carson, R. 2002. *Silent spring.* 40th anniversary ed., 1st Mariner Books ed edition. Boston, Houghton Mifflin. 378 pp.

Carvalho, F. P. 2017. Pesticides, environment, andfoodsafety. *FoodandEnergySecurity*, 6(2): 48–60. https: //doi. org/10. 1002/fes3. 108.

CCICED. 2015. SpecialPolicyStudyonSoilPollutionManagement. ChinaCouncil for International Cooperation on Environment and Development. (also available at http: //english. sepa. gov. cn/ Events/Special_Topics/AGM_1/2015nh/document/201605/P020160524149463335883. pdf).

CDC. 2013. AntibioticResistanceThreatsintheUnitedStates. U. S. Departmentof Helath and Human Services, Centers for Disease Control and Prevention. (also available at https: //www. cdc. gov/ drugresistance/pdf/ar-threats-2013-508. pdf).

Cernansky, R. 2015. Agriculture: State-of-the-art soil. *Nature News*, 517(7534): 258. https: //doi. org/10. 1038/517258a.

Certini, G., Scalenghe, R. &Woods, W. I. 2013. Theimpactofwarfareonthe soil environment. *Earth-Science Reviews*, 127: 1–15. https: //doi. org/10. 1016/j. earscirev. 2013. 08. 009.

Cestti, R., Srivastava, J. P. &Jung, S. 2003. *AgricultureNon-PointSourcePollutionControl.* World Bank Working Papers. The World Bank. 54 pp. (also available at https: // elibrary. worldbank.

org/doi/abs/10. 1596/0-8213-5523-6).

Cetin, B. 2016. InvestigationofPAHs, PCBsandPCNsinsoilsaroundaHeavily IndustrializedAreainKocaeli, Turkey: Concentrations, distributions, sourcesand toxicologicaleffects. *ScienceofTheTotalEnvironment*, 560–561: 160–169. https: //doi. org/10. 1016/j. scitotenv. 2016. 04. 037.

CFR. 2017. Sec. 178. 3740 Plasticizers in polymeric substances. [Cited 3 April 2018]. https: //www. accessdata. fda. gov/scripts/cdrh/cfdocs/cfcfr/CFRSearch. cfm?fr=178. 3740.

CGIAR&CCAFS. 2018. IntegratedSoilFertilityManagement(ISFM)|Climate-Smart Agriculture Guide. In: *CLIMATE-SMART AGRICULTURE* [online]. [Cited 3 April 2018]. https: //csa. guide/csa/integrated-soil-fertility-management-isfm.

Chalew, T. E. A. &Halden, R. U. 2009. EnvironmentalExposureofAquaticandTerrestrial Biota to Triclosan and Triclocarban. *JAWRA Journal of the American Water Resources Association*, 45(1): 4–13. https: //doi. org/10. 1111/j. 1752-1688. 2008. 00284. x.

Chaney, R. L. 1980. Health risks associated with toxic metals in municipal sludges. *In* G. Bitton, B. L. Damron, G. T. Edds& J. M. (eds) Davidson, eds. *Sludge: health risks of land application.*, pp. 59–83. Ann Arbor, MI. (also available at https: //www. osti. gov/ biblio/6671808).

Chaney, R. L. &Hornick, S. B. 1977. Accumulationandeffectsofcadmiumoncrops. *Cadmium 77*. pp. 125–140. Paper presented at Proceedings of the First International Cadmium Conference, 1977, SanFrancisco.

ChaparroLeal, L. T., Guney, M. &Zagury, G. J. 2018. Invitrodermalbioaccessibilityof selected metals in contaminated soil and mine tailings and human health risk characterization. *Chemosphere*, 197: 42–49. https: //doi. org/10. 1016/j. chemosphere. 2018. 01. 008.

Chaudhry, Q., Blom-Zandstra, M., Gupta, S. K. &Joner, E. 2005. UtilisingtheSynergy between Plants and Rhizosphere Microorganisms to Enhance Breakdown of Organic Pollutants in the Environment (15 pp). *Environmental Science* and *Pollution Research-International*, 12(1): 34–48. https: //doi. org/10. 1065/espr2004. 08. 213.

Chen, C., Wang, Y., Qian, Y., Zhao, X. &Wang, Q. 2015. Thesynergistictoxicityofthe multiple chemical mixtures: Implications for risk assessment in the terrestrial environment. *Environment International*, 77: 95–105. https: //doi. org/10. 1016/j. envint. 2015. 01. 014.

Chen, Y., Li, X. &Shen, Z. 2004. Leachinganduptakeofheavymetalsbytendifferent species of plants during an EDTA-assisted phytoextraction process. *Chemosphere*, 57(3): 187–196. https: //doi. org/10. 1016/j. chemosphere. 2004. 05. 044.

Civan, A., Worrall, F., Jarvie, H. P., Howden, N. J. K. &Burt, T. P. 2018. Forty-yeartrendsin the flux and concentration of phosphorus in British rivers. *Journal of Hydrology*, 558: 314–327. https: //doi. org/10. 1016/j. jhydrol. 2018. 01. 046.

Clarke, J. M., Leisle, D., DePauw, R. M. & Thiessen, L. L. 1997. Registration of Five Pairs of Durum Wheat Genetic Stocks Near-Isogenic for Cadmium Concentration. *Crop Science*, 37(1): 297. https: //doi. org/10. 2135/cropsci1997. 0011183X003700010071x.

Cole, K. J. 2015. *Bacterial Counts In Composted* and *Fresh Recycled Dairy Manure Bedding*. The Ohio State University. (also available at https: //etd. ohiolink. edu/ pg_10?0: : NO: 10: P10_ACCESSION_NUM: osu1429188763).

Collins, C., Fryer, M. & Grosso, A. 2006. Plant Uptake of Non-Ionic Organic Chemicals. *Environmental Science & Technology*, 40(1): 45–52. https: //doi. org/10. 1021/es0508166.

Committee on Bioavailability of Contaminants in Soils and Sediments. 2002. *Bioavailability of Contaminants in Soils* and *Sediments: Processes, Tools,* and *Applications*. Washington, National Research Council of the National Academies. (also available at https: // www. nap. edu/catalog/10523/bioavailability-of-contaminants-in-soils-and-sediments-processes-tools-and).

Conselho Nacional do Meio Ambiente. 2009. *RESOLUÇÃO No 420, DE 28 DE DEZEMBRO DE 2009. Dispõe sobre critérios e valores orientadores de qualidade do solo quanto à presença de substâncias químicas e estabelece diretrizes para o gerenciamento ambiental de áreas contaminadas por essas substâncias em decorrência de atividades antrópicas*. [online]. [Cited 3 April 2018]. http: //www. mma. gov. br/port/conama/res/res09/res42009. pdf.

Contaminated Sites Management Working Group. 1999. A Federal approach to contaminated sites. Contaminated Sites Management Working Group (CSMWG).

Conte, P., Zena, A., Pilidis, G. & Piccolo, A. 2001. Increased retention of polycyclic aromatic hydrocarbons in soils induced by soil treatment with humic substances. *Environmental Pollution*, 112(1): 27–31. https: //doi. org/10. 1016/S0269-7491(00)00101-9.

Covaci, A., Geens, T., Roosens, L., Ali, N., VandenEede, N., Ionas, A. C., Malarvannan, G. & Dirtu, A. C. 2011. Human Exposure and Health Risks to Emerging Organic Contaminants. *In* D. Barceló, ed. *Emerging Organic Contaminants* and *Human Health*, pp. 243–305. Berlin, Heidelberg, Springer Berlin Heidelberg. (also available at http: //link. springer. com/10. 1007/698_2011_126).

Craig, Z. R., Wang, W. & Flaws, J. A. 2011. Endocrine-disrupting chemicals in ovarian function: effects on steroidogenesis, metabolism and nuclear receptor signaling. *Reproduction (Cambridge, England)*, 142(5): 633–646. https: //doi. org/10. 1530/REP-11- 0136.

Cruz, N., Rodrigues, S. M., Coelho, C., Carvalho, L., Duarte, A. C., Pereira, E. & Römkens, P. F. A. M. 2014. Urban agriculture in Portugal: Availability of potentially toxic elements for plant uptake. *Applied Geochemistry*, 44: 27–37. https: //doi. org/10. 1016/j. apgeochem. 2013. 07. 003 Ćujić, M., Dragović, S., Đorđević, M., Dragović, R., Gajić, B. & Miljanić, Š. 2015. Radionuclides in the soil around the largest coal-fired power plant in Serbia: radiological hazard, relationship with soil characteristics and spatial distribution. *Environmental Science* and *Pollution Research*, 22(13): 10317–10330. https: //doi. org/10. 1007/s11356-014-3888-2.

Cytryn, E. 2013. The soil resistome: The anthropogenic, the native, and the unknown. *Soil Biology* and *Biochemistry*, 63: 18–23. https: //doi. org/10. 1016/j. soilbio. 2013. 03. 017.

Dalkmann, P., Siebe, C., Amelung, W., Schloter, M. & Siemens, J. 2014. Does Long-Term Irrigation with Untreated Wastewater Accelerate the Dissipation of Pharmaceuticals in Soil? *Environmental Science & Technology*, 48(9): 4963–4970. https: //doi. org/10. 1021/ es501180x.

Darnerud, P. O. 2003. Toxic effects of brominated flame retardants in man and in wildlife. *Environment International*, 29(6): 841–853. https: //doi. org/10. 1016/S0160- 4120(03)00107-7.

Daughton, C. G. &Ternes, T. A. 1999. Pharmaceuticals and personal care products in the environment: agents of subtle change? *Environmental Health Perspectives*, 107 Suppl 6: 907–938.

DEA. 2010. Framework for the Management of Contaminated Land. Republic of South Africa, Department of Environmental Affairs. (also available at http: //sawic. environment. gov. za/ documents/562. pdf).

Deardorff, T., Karch, N. &Holm, S. 2008. Dioxin levels in ash and soil generated in Southern California fires. *Organohalogen Compounds*, 70: 2284–2288.

DECA. 2010. Assessment Levels for Soil, Sediment and Water., p. 56. No. 4. Australia, Department of Environment and Conservation. (also available at https: // www. der. wa. gov. au/ images/documents/your-environment/contaminated-sites/guidelines/2009641_-_assessment_ levels_for_soil_sediment_and_water_-_web. pdf).

Diaconu, A., enu, I., Roca, R. &Cârlescu, P. 2017. Researches regarding the reduction of pesticide soil pollution in vineyards. *Process Safety* and *Environmental Protection*, 108: 135–143. https: // doi. org/10. 1016/j. psep. 2016. 09. 016.

Díaz-Cruz, M. S. &Barceló, D. 2005. LC–MS2 trace analysis of antimicrobials in water, sediment and soil. *TrAC Trends in Analytical Chemistry*, 24(7): 645–657. https: //doi. org/10. 1016/j. trac. 2005. 05. 005.

Díez, M., Simón, M., Martín, F., Dorronsoro, C., García, I. &VanGestel, C. A. M. 2009. Ambient trace element background concentrations in soils and their use in risk assessment. *Science of The Total Environment*, 407(16): 4622–4632. https: //doi. org/10. 1016/j. scitotenv. 2009. 05. 012.

Ding, G. -C., Radl, V., Schloter-Hai, B., Jechalke, S., Heuer, H., Smalla, K. &Schloter, M. 2014. Dynamics of Soil Bacterial Communities in Response to Repeated Application of Manure Containing Sulfadiazine. *PLoS ONE*, 9(3): e92958. https: //doi. org/10. 1371/ journal. pone. 0092958.

Dœlsch, E., SaintMacary, H. & VandeKerchove, V. 2006. Sources of very high heavy metal content in soils of volcanic island (La Réunion). *Journal of Geochemical Exploration*, 88(1–3): 194–197. https://doi.org/10.1016/j.gexplo.2005.08.037.

Doran, J. W., Stamatiadis, S. & Haberern, J. 2002. Soil health as an indicator of sustainable management. *Publications from USDA-ARS / UNL Faculty*. (also available at https://digitalcommons.unl.edu/usdaarsfacpub/180).

Dores, E. F. G. C., Spadotto, C. A., Weber, O. L. S., DallaVilla, R., Vecchiato, A. B. & Pinto, A. A. 2016. Environmental Behavior of Chlorpyrifos and Endosulfan in a Tropical Soil in Central Brazil. *Journal of Agricultural* and *Food Chemistry*, 64(20): 3942–3948. https://doi.org/10.1021/acs.jafc.5b04508.

Dorta-Santos, M., Tejedor, M., Jiménez, C., Hernández-Moreno, J., Palacios-Díaz, M. & Díaz, F. 2014. Recycled Urban Wastewater for Irrigation of Jatropha curcas L. in Abandoned Agricultural Arid Land. *Sustainability*, 6(10): 6902–6924. https://doi.org/10.3390/su6106902.

Du, L. & Liu, W. 2012. Occurrence, fate, and ecotoxicity of antibiotics in agro-ecosystems. A review. *Agronomy for Sustainable Development*, 32(2): 309–327. https://doi.org/10.1007/s13593-011-0062-9.

Dubois, O. 2011. *The State of the World's Land and Water Resources for Food and Agriculture: Managing Systems at Risk*. Routledge. (also available at https://www.taylorfrancis.com/books/9780203142837).

EA, ed. 2008. *Updated technical background to the CLEA model: using science to create a better place*. Bristol, Environment Agency of Great Britain. 155pp.

EC. 1986. Council Directive 86/278/EEC of 12 June 1986 on the protection of the environment, and in particular of the soil, when sewage sludge is used in agriculture. [Cited 3 April 2018]. https://eur-lex.europa.eu/legal-content/EN/TXT/HTML/?uri=CELEX:31986L0278&from=EN.

EC. 1991. Council Directive 91/676/EEC of 12 December 1991 concerning the protection of waters against pollution caused by nitrates from agricultural sources. [Cited 3 April 2018]. http://eur-lex.europa.eu/legal-content/EN/TXT/HTML/?uri=CELEX:31991L0676&from=EN.

EC. 1996. Council Directive 96/61/EC of 24 September 1996 concerning integrated pollution prevention and control. [Cited 3 April 2018]. http://eur-lex.europa.eu/LexUriServ/LexUriServ.do?uri=CELEX:31996L0061:en:HTML.

EC. 2002. Integrated Crop Management Systems in the EU., p. 157. No. 1882/BDB/ May 2002. European Commission DG Environment.

EC. 2006. REGULATION (EC) No 1907/2006 concerning the Registration, Evaluation, Authorisation and Restriction of Chemicals (REACH), establishing a European Chemicals Agency, amending Directive 1999/45/EC and repealing Council Regulation (EEC)

No793/93. andCommissionRegulation(EC)No1488/94 as well as Council Directive 76/769/EEC and Commission Directives 91/155/EEC, 93/67/EEC, 93/105/ECand2000/21/EC. [Cited3April2018]. https: //eur-lex. europa. eu/legal-content/EN/TXT/HTML/?uri=CELEX: 02006R1907-20140410&from=EN.

EC. 2011. CommissionRegulationNo835/2011. amendingRegulation(EC)No1881/2006 asregardsmaximumlevelsforpolycyclicaromatic. [Cited3April2018]. https: //eur- lex. europa. eu/legal-content/EN/TXT/HTML/?uri=CELEX: 32011R0835&from=EN.

EC. 2015. COM/2015/0614final. Closingtheloop-AnEUactionplanfortheCircular Economy. [Cited 3 April 2018]. https: //eur-lex. europa. eu/legal-content/EN/TXT/ HTML/?uri=CELEX: 52015DC0614&from=EN.

EC. 2017. A European One Health Action Plan against Antimicrobial Resistance (AMR). European Commission. (also available at https: //ec. europa. eu/health/amr/ sites/amr/files/amr_action_plan_2017_en. pdf).

EEA. 2014. Progressinmanagementofcontaminatedsites. EuropeanEnvironment Agency. (also available at https: //www. eea. europa. eu/data-and-maps/indicators/progress-in-management-of-contaminated-sites/progress-in-management-of- contaminated-1).

EFSA. 2008. Scientific opinion of the panel on contaminants in the food chain on polycyclic aromatic hydrocarbons in food. *The EFSA Journal*, 724: 1–114.

Ensink, J. H. J., Mahmood, T., Hoek, W. vander, Raschid-Sally, L. &Amerasinghe, F. P. 2004. A nationwide assessment of wastewater use in Pakistan: an obscure activity or a vitally important one? *Water Policy*, 6(3): 197–206.

Ercumen, A., Pickering, A. J., Kwong, L. H., Arnold, B. F., Parvez, S. M., Alam, M., Sen, D., Islam, S., Kullmann, C., Chase, C., Ahmed, R., Unicomb, L., Luby, S. P. &Colford, J. M. 2017. Animal Feces Contribute to Domestic Fecal Contamination: Evidence from *E. coli* Measured in Water, Hands, Food, Flies, and Soil in Bangladesh. *Environmental Science&Technology*, 51(15): 8725–8734. https: //doi. org/10. 1021/acs. est. 7b01710.

EuropeanCommission, JointResearchCentre&GlobalSoilBiodiversityInitiative. 2016. *Global soil diversity atlas*. (also available at http: //esdac. jrc. ec. europa. eu/public_path/JRC_global_soilbio_atlas_online. pdf).

EUROSTAT. 2018. *Chemicalsproductionandconsumptionstatistics-StatisticsExplained* [online]. [Cited 3 April 2018]. http: //ec. europa. eu/eurostat/statistics-explained/index. php/Chemicals_production_and_consumption_statistics.

Fabiańska, M. J., Kozielska, B., Konieczyński, J. & Kowalski, A. 2016. Sources of organic pollution in particulate matter and soil of Silesian Agglomeration (Poland): evidence from geochemical markers. *Environmental Geochemistry* and *Health*, 38(3): 821–842. https: //doi.

org/10. 1007/s10653-015-9764-2.

Fabra, A. 1997. Toxicity of 2, 4-Dichlorophenoxyacetic Acid to Rhizobium sp in Pure Culture. *Bulletin of Environmental Contamination* and *Toxicology*, 59(4): 645–652.

Falciglia, P. P., Cannata, S., Romano, S. & Vagliasindi, F. G. 2014. Stabilisation/solidification of radionuclide polluted soils - Part I: Assessment of setting time, mechanical resistance, g-radiation shielding and leachate g-radiation. *JOURNAL OF GEOCHEMICAL EXPLORATION*, 142: 104–111.

FAO. 2000. Assessing soil contamination A reference manual. Rome, Italy, Food and Agriculture Organization of the United Nations. (also available at http: //www. fao. org/docrep/003/X2570E/X2570E00. HTM).

FAO. 2003. International Code of Conduct on the Distribution and Use of Pesticides. Rome, Italy, Food and Agriculture Organization of the United Nations. (also available at http: //www. fao. org/docrep/005/Y4544E/y4544e00. htm).

FAO. 2006. International Code of Conduct on the Distribution and Use of Pesticides. Guidelines on Efficacy Evaluation for the Registration of Plant Protection Products. Rome, Italy, Food and Agriculture Organization of the United Nations. (also available at http: //www. fao. org/fileadmin/templates/agphome/documents/Pests_Pesticides/Code/Efficacy. pdf).

FAO. 2015a. *World fertilizer trends* and *outlook to 2018*. Rome, Food and Agriculture Organization of the United Nations Statistics. (also available at http: //www. fao. org/3/a-i4324e. pdf).

FAO. 2015b. World Soil Charter., p. 10. Rome, Italy, Food and Agriculture Organization of the United Nations. (also available at http: //www. fao. org/3/a-i4965e. pdf).

FAO. 2015c. *FAO statistical pocketbook 2015: world food* and *agriculture*. Food and Agriculture Organization of the United Nations Statistics.

FAO. 2016. The FAO Action Plan on Antimicrobial Resistance 2016-2020., p. 25. Rome, Italy, Food and Agriculture Organization of the United Nations. (also available at http: //www. fao. org/3/a-i5996e. pdf).

FAO. 2017. Voluntary Guidelines for Sustainable Soil Management. Rome, Italy, Food and Agriculture Organization of the United Nations. (also available at http: // www. fao. org/3/a-bl813e. pdf).

FAO & ITPS. 2015. Status of the World's Soil Resources (SWSR) - Main Report. Rome, Italy, ood and Agriculture Organization of the United Nations and Intergovernmental Technical Panel on Soils. (also available at http: //www. fao. org/3/a-i5199e. pdf).

FAO & ITPS. 2017. Global assessment of the impact of plant protection products on soil functions and soil ecosystems. Rome, Italy, Food and Agriculture Organization of the United Nations. (also available at http: //www. fao. org/3/I8168EN/i8168en. pdf).

FAO&WHO, eds. 2016. *InternationalCodeofConductonPesticideManagement. Guidelines onHighlyHazardousPesticides.* HealthandsafetyguideNo. 41. Rome, Italy, Foodand Agriculture-OrganizationoftheUnitedNations. 24pp.

FAOSTAT. 2016. *FAOSTATInputs/PesticidesUse.* [online]. [Cited3April2018]. http: // www. fao. org/faostat/en/#data/RP.

Farenhorst, A., Papiernik, S. K., Saiyed, I., Messing, P., Stephens, K. D., Schumacher, J. A., Lobb, D. A., Li, S., Lindstrom, M. J. &Schumacher, T. E. 2008. HerbicideSorptionCoefficientsin Relation to Soil Properties and Terrain Attributes on a Cultivated Prairie. *Journal of EnvironmentQuality*, 37(3): 1201. https: //doi. org/10. 2134/jeq2007. 0109.

Fesenko, S., Howard, B. J., Sanzharova, N. I. & Vidal, M. 2017. Remediation of areas contaminated by caesium: Basic mechanisms behind remedial options and experience in application. *In* D. K. Gupta & C. Walther, eds. *Impact of Cesium on Plants* and *the Environment*, pp. 265–310. Springer International Publishing. (also available at//www. springer. com/gp/book/9783319415246).

Fesenko, S. V., Alexakhin, R. M., Balonov, M. I., Bogdevitch, I. M., Howard, B. J., Kashparov, V. A., Sanzharova, N. I., Panov, A. V., Voigt, G. &Zhuchenka, Y. M. 2007. An extendedcriticalreview of twenty years of countermeasures used in agriculture after the Chernobyl accident. *Science of The Total Environment*, 383(1): 1–24. https: //doi. org/10. 1016/j. scitotenv. 2007. 05. 011.

Fiedler, H., Abad, E., vanBavel, B., deBoer, J., Bogdal, C. &Malisch, R. 2013. Theneedfor capacity building and first results for the Stockholm Convention Global Monitoring Plan. *TrACTrendsin Analytical Chemistry*, 46: 72–84. https: //doi. org/10. 1016/j. trac. 2013. 01. 010.

Fiorino, D. J. 2010. Voluntary initiatives, regulations, and nanotechnology oversight: Charting a Path. Woodrow Wilson International Center for Scholars & The Pew ChariTableTrusTs. (also available at http: //www. nanotechproject. org/process/ assets/files/8347/pen-19. pdf).

Flores-Magdaleno, H., Mancilla-Villa, O. R., Mejía-Saenz, E., Olmedo-Bolaños, M. C. &Bautista-Olivas, A. L. 2011. HeavymetalsInagriculturalsoilsandIrrigationwastewater of Mixquiahuala, Hidalgo, Mexico. *African Journal of Agricultural Research*, 6(24). https: //doi. org/10. 5897/AJAR11. 414.

FOEN. 2013. *Fundamental approach* [online]. [Cited 3 April 2018]. https: //www. bafu. admin. ch/bafu/en/home/topics/contaminated-sites/info-specialists/remediation- of-contaminated-sites/fundamental-approach. html.

Fox, J. E., Gulledge, J., Engelhaupt, E., Burow, M. E. &McLachlan, J. A. 2007. Pesticides reduce symbiotic efficiency of nitrogen-fixing rhizobia and host plants. *Proceedings of the National Academy of Sciences*, 104(24): 10282–10287. https: //doi. org/10. 1073/ pnas. 0611710104.

Fritt-Rasmussen, J., Jensen, P. E., Christensen, R. H. B. &Dahllöf, I. 2012. Hydrocarbonand Toxic

Metal Contamination from Tank Installations in a Northwest Greenlandic Village. *Water, Air, &SoilPollution*, 223(7): 4407–4416. https: //doi. org/10. 1007/s11270- 012-1204-7.

Frumin, G. T. &Gildeeva, I. M. 2014. Eutrophicationof water bodies—Aglobalenvironmental problem. *Russian Journal of General Chemistry*, 84(13): 2483–2488. https: //doi. org/10. 1134/ S1070363214130015.

García, C. &Lobo, M. 2013. Rehabilitacióndesuelosdegradadosycontaminadosmediante la aplicación de compost. *In* J. M. Casco, ed. *Compostaje*, pp. 425–448. Madrid, Spain, Mundi-Prensa. (also available at https: //library. biblioboard. com/ content/ef8fbfdb-d094-4eea-b6f0-3be6e903c232).

Garcia, R., Baelum, J., Fredslund, L., Santorum, P. & Jacobsen, C. S. 2010. Influence of Temperature and Predation on Survival of Salmonella enterica Serovar Typhimurium and Expression of invAin Soil and Manure-Amended Soil. *Applied andEnvironmentalMicrobiology*, 76(15): 5025–5031. https: //doi. org/10. 1128/AEM. 00628- 10.

García-Pérez, J., Boldo, E., Ramis, R., Pollán, M., Pérez-Gómez, B., Aragonés, N. &López-Abente, G. 2007. DescriptionofindustrialpollutioninSpain. *BMCPublicHealth*, 7(1). https: //doi. org/10. 1186/1471-2458-7-40.

García-Préchac, F., Ernst, O., Siri-Prieto, G. &Terra, J. A. 2004. Integratingno-tillinto crop–pasture rotations in Uruguay. *Soil* and *Tillage Research*, 77(1): 1–13. https: //doi. org/10. 1016/j. still. 2003. 12. 002.

Geissen, V., Mol, H., Klumpp, E., Umlauf, G., Nadal, M., vanderPloeg, M., vandeZee, S. E. A. T. M. &Ritsema, C. J. 2015. Emergingpollutantsintheenvironment: Achallengeforwater resource management. *International Soil* and *Water Conservation Research*, 3(1): 57–65. https: //doi. org/10. 1016/j. iswcr. 2015. 03. 002.

Gevao, B., Semple, K. T. &Jones, K. C. 2000. Boundpesticideresiduesinsoils: areview. *Environmental Pollution (Barking, Essex: 1987)*, 108(1): 3–14.

Ghisari, M. &Bonefeld-Jorgensen, E. C. 2009. Effectsofplasticizersandtheirmixtures on estrogen receptor and thyroid hormone functions. *Toxicology Letters*, 189(1): 67– 77. https: //doi. org/10. 1016/j. toxlet. 2009. 05. 004.

Giesy, J. P. &Kannan, K. 2001. GlobalDistributionofPerfluorooctaneSulfonatein Wildlife. *Environmental Science &Technology*, 35(7): 1339–1342. https: //doi. org/10. 1021/ es001834k.

Glick, B. R. 2003. Phytoremediation: synergisticuseofplantsandbacteriatoclean uptheenvironment. *BiotechnologyAdvances*, 21(5): 383–393. https: //doi. org/10. 1016/ S0734-9750(03)00055-7.

Godfray, H. C. J., Beddington, J. R., Crute, I. R., Haddad, L., Lawrence, D., Muir, J. F., Pretty, J., Robinson, S., Thomas, S. M. &Toulmin, C. 2010. FoodSecurity: TheChallengeof Feeding 9BillionPeople. *Science*, 327(5967): 812–818. https: //doi. org/10. 1126/science.

1185383.

Gomes, H. I., Dias-Ferreira, C. &Ribeiro, A. B. 2013. Overviewofinsituandexsitu remediationtechnologiesforPCB-contaminatedsoilsandsedimentsandobstacles forfull-scaleapplication. *ScienceofTheTotalEnvironment*, 445–446: 237–260. https://doi.org/10.1016/j.scitotenv.2012.11.098.

González-Naranjo, V. &Boltes, K. 2014. Toxicityofibuprofenandperfluorooctanoic acid for risk assessment of mixtures in aquatic and terrestrial environments. *International Journal of Environmental Science* and *Technology*, 11(6): 1743–1750. https://doi.org/10.1007/s13762-013-0379-9.

González-Naranjo, V., Boltes, K., deBustamante, I. &Palacios-Diaz, P. 2015. Environmental risk of combined emerging pollutants in terrestrial environments: chlorophyll a fluorescence analysis. *Environmental Science* and *Pollution Research*, 22(9): 6920–6931. https://doi.org/10.1007/s11356-014-3899-z.

Good, A. G. &Beatty, P. H. 2011. FertilizingNature: ATragedyofExcessintheCommons. *PLoS Biology*, 9(8): e1001124. https://doi.org/10.1371/journal.pbio.1001124.

Gottesfeld, P., Were, F. H., Adogame, L., Gharbi, S., San, D., Nota, M. M. &Kuepouo, G. 2018. Soil contamination from lead battery manufacturing and recycling in sevenAfrican countries. *Environmental Research*, 161: 609–614. https://doi.org/10.1016/j.envres.2017.11.055.

Gottschall, N., Topp, E., Metcalfe, C., Edwards, M., Payne, M., Kleywegt, S., Russell, P. &Lapen, D. R. 2012. Pharmaceutical and personal care products in groundwater, subsurface drainage, soil, and wheat grain, following a high single application of municipal biosolids to a field. *Chemosphere*, 87(2): 194–203. https://doi.org/10.1016/j.chemosphere.2011.12.018.

Gotz, A. &Smalla, K. 1997. ManureEnhancesPlasmidMobilizationandSurvival of Pseudomonas putida Introduced into Field Soil. *Applied* and *Environmental Microbiology*, 63(5): 1980–1986.

vanderGraaf, E. R., Koomans, R. L., Limburg, J. &deVries, K. 2007. Insituradiometric mapping as a proxy of sediment contamination: Assessment of the underlying geochemical and-physical principles. *Applied Radiation* and *Isotopes*, 65(5): 619–633. https://doi.org/10.1016/j.apradiso.2006.11.004.

Grant, C. A., Bailey, L. D., McLaughlin, M. & Singh, B. R. 1999. Management factors which inflience cadmium concentrations in crops. *In* M. J. McLaughlin & B. R. Singh, eds. *Cadmium in Soils* and *Plants*, pp. 151–198. Developments in Plant and Soil Sciences. Springer Netherlands. (also available at //www.springer.com/gp/book/9780792358435).

Grant, C. A., Clarke, J. M., Duguid, S. &Chaney, R. L. 2008. Selectionandbreedingofplant cultivars to minimize cadmium accumulation. *The Science of the Total Environment*, 390(2–3): 301–310. https://doi.org/10.1016/j.scitotenv.2007.10.038.

Grass, G., Rensing, C. &Solioz, M. 2011. MetallicCopperasanAntimicrobialSurface. *Applied and Environmental Microbiology*, 77(5): 1541–1547. https: //doi. org/10. 1128/ AEM. 02766-10.

Grathwohl, P. & Halm, D., eds. 2003. INTEGRATED SOIL and WATER PROTECTION: RISKSFROMDIFFUSEPOLLUTION. *Clustermeeting*；*2nd, Innovativemanagement of groundwater resources in Europe - training* and *RTD coordination*；*Sustainable management of soil* and *groundwater resources in urban areas*. Conference papers /Umweltbundesamt, Wien. Paperpresentedat, 2003, Wien.

Gregorič, A., Vaupotič, J., Kardos, R., Horváth, M., Bujtor, T. &Kovács, T. 2013. Radon emanation of soils from different lithological units. *Carpathian Journal of Earth and Environmental Sciences*, 8(2): 185–190.

Grobelak, A. 2016. Organic Soil Amendments in the Phytoremediation Process. *In* A. A. Ansari, S. S. Gill, R. Gill, G. R. Lanza & L. Newman, eds. *Phytoremediation*, pp. 21–39. Cham, Springer International Publishing. (also available at http: //link. springer. com/10. 1007/978-3-319-41811-7_2).

Grobelak, A. &Napora, A. 2015. TheChemophytostabilisationProcessofHeavy Metal Polluted Soil. *PLOS ONE*, 10(6): e0129538. https: //doi. org/10. 1371/journal. pone. 0129538.

Groot, R. S. de. 1992. *Functionsofnature: evaluationofnatureinenvironmentalplanning. management* and *decision making*. Groningen, Wolters-Noordhoff. 315pp.

GSP. 2017. ReportoftheFifthMeetingofthePlenaryAssembly(PA)oftheGlobal Soil Partnership (GSP). Rome, Italy, Food and Agriculture Organization of the UnitedNations. (alsoavailableathttp: //www. fao. org/3/a-bs973e. pdf).

Guerra, F., Trevizam, A. R., Muraoka, T., Marcante, N. C. &Canniatti-Brazaca, S. G. 2012. Heavy metals in vegetables and potential risk for human health. *Scientia Agricola*, 69(1): 54–60. https: //doi. org/10. 1590/S0103-90162012000100008.

Guillén, J., Muñoz-Muñoz, G., Baeza, A., Salas, A. &Mocanu, N. 2017. Modificationofthe137Cs, 90Sr, and 60Co transfer to wheat plantlets by NH4+ fertilizers. *Environmental science and pollution research international*, 24(8): 7383–7391. https: //doi. org/10. 1007/ s11356-017-8439-1.

Gulkowska, A., Sander, M., Hollender, J. &Krauss, M. 2013. CovalentBindingof Sulfamethazine to Natural and Synthetic Humic Acids: Assessing Laccase Catalysis and Covalent Bond Stability. *Environmental Science &Technology*, 47(13): 6916–6924. https: //doi. org/10. 1021/ es3044592.

Gullberg, E., Cao, S., Berg, O. G., Ilbäck, C., Sandegren, L., Hughes, D. &Andersson, D. I. 2011. Selection of Resistant Bacteria at Very Low Antibiotic Concentrations. *PLoS Pathogens*, 7(7): e1002158. https: //doi. org/10. 1371/journal. ppat. 1002158.

Guo, J. H., Liu, X. J., Zhang, Y., Shen, J. L., Han, W. X., Zhang, W. F., Christie, P., Goulding, K. W. T., Vitousek, P. M. & Zhang, F. S. 2010. Significant Acidification in Major Chinese Croplands. *Science*, 327(5968): 1008–1010. https: //doi. org/10. 1126/science. 1182570.

Guo, K., Liu, Y. F., Zeng, C., Chen, Y. Y. &Wei, X. J. 2014. Globalresearchonsoilcontamination from1999to2012: Abibliometricanalysis. *ActaAgriculturaeScandinavica, SectionB — Soil & Plant Science*, 64(5): 377–391. https: //doi. org/10. 1080/09064710. 2014. 913679.

Guzzella, L., Poma, G., De Paolis, A., Roscioli, C. &Viviano, G. 2011. Organic persistent toxic substances in soils, waters and sediments along an altitudinal gradient at Mt. Sagarmatha, Himalayas, Nepal. *Environmental Pollution*, 159(10): 2552–2564. https: // doi. org/10. 1016/j. envpol. 2011. 06. 015.

Hafez, Y. I. &Awad, E. -S. 2016. Finiteelementmodelingofradondistributionin natural soils of different geophysical regions. *Cogent Physics*, 3(1). https: //doi. org/10. 1080/23311940. 2016. 1254859.

Halling-Sørensen, B., Jensen, J., Tjørnelund, J. &Montforts, M. H. M. M. 2001. Worst-Case Estimations of Predicted Environmental Soil Concentrations (PEC) of Selected Veterinary Antibiotics and Residues Used in Danish Agriculture. *In* K. Kümmerer, ed. *Pharmaceuticals in the Environment*, pp. 143–157. Berlin, Heidelberg, Springer Berlin Heidelberg. (also available at http: //link. springer. com/10. 1007/978-3-662- 04634-0_13).

Halling-Sørensen, B., NorsNielsen, S., Lanzky, P. F., Ingerslev, F., HoltenLützhøft, H. C. &Jørgensen, S. E. 1998. Occurrence, fateandeffectsofpharmaceuticalsubstances in the environment- A review. *Chemosphere*, 36(2): 357–393. https: //doi. org/10. 1016/ S0045-6535(97)00354-8.

Hamon, R. E., Lombi, E., Fortunati, P., Nolan, A. L. & McLaughlin, M. J. 2004. Coupling Speciationand Isotope Dilution Techniques To Study Arsenic Mobilization in the Environment. *Environmental Science &Technology*, 38(6): 1794–1798. https: //doi. org/10. 1021/es034931x.

Hamscher, G., Pawelzick, H. T., Höper, H. &Nau, H. 2004. AntibioticsinSoil: Routes of Entry, Environmental Concentrations, Fate and Possible Effects. *In* K. Kümmerer, ed. *Pharmaceuticals in the Environment*, pp. 139–147. Berlin, Heidelberg, Springer Berlin Heidelberg. (also available at http: //www. springerlink. com/ index/10. 1007/978-3-662-09259-0_11).

Han, J., Shi, J., Zeng, L., Xu, J. &Wu, L. 2015. Effectsofnitrogenfertilizationonthe acidity and salinity of greenhouse soils. *Environmental Science* and *Pollution Research*, 22(4): 2976–2986. https: //doi. org/10. 1007/s11356-014-3542-z.

Handy, R. D., vonderKammer, F., Lead, J. R., Hassellöv, M., Owen, R. &Crane, M. 2008. The ecotoxicology and chemistry of manufactured nanoparticles. *Ecotoxicology*, 17(4): 287–314.

https://doi.org/10.1007/s10646-008-0199-8.

Hao, X., Chang, C., Travis, G. R. & Zhang, F. 2003. Soil carbon and nitrogen response to 25 annual cattle manure applications. *Journal of Plant Nutrition* and *Soil Science*, 166(2): 239–245. https://doi.org/10.1002/jpln.200390035.

Harbarth, S., Balkhy, H. H., Goossens, H., Jarlier, V., Kluytmans, J., Laxminarayan, R., Saam, M., VanBelkum, A. & Pittet, D. 2015. Antimicrobial resistance: one world, one fight! *Antimicrobial Resistance* and *Infection Control*, 4(1). https://doi.org/10.1186/s13756-015-0091-2.

Hargreaves, J., Adl, M. & Warman, P. 2008. A review of the use of composted municipal solid waste in agriculture. *Agriculture, Ecosystems & Environment*, 123(1–3): 1–14. https://doi.org/10.1016/j.agee.2007.07.004.

Hashim, T., Abbas, H., Farid, I., El-Husseiny, O. & Abbas, M. 2017. Accumulation of some heavy metals in plants and soils adjacent to Cairo – Alexandria agricultural highway. *Egyptian Journal of Soil Science*, 0(0): 0–0. https://doi.org/10.21608/ejss.2016.281.1047.

Heberer, T. 2002. Occurrence, fate, and removal of pharmaceutical residues in the aquatic environment: a review of recent research data. *Toxicology Letters*, 131(1–2): 5–17.

Hernández, A. F., Parrón, T., Tsatsakis, A. M., Requena, M., Alarcón, R. & López-Guarnido, O. 2013. Toxic effects of pesticide mixtures at a molecular level: Their relevance to human health. *Toxicology*, 307: 136–145. https://doi.org/10.1016/j.tox.2012.06.009.

Heudorf, U., Mersch-Sundermann, V. & Angerer, J. 2007. Phthalates: toxicology and exposure. *International Journal of Hygiene* and *Environmental Health*, 210(5): 623–634. https://doi.org/10.1016/j.ijheh.2007.07.011.

Heuer, H., Focks, A., Lamshöft, M., Smalla, K., Matthies, M. & Spiteller, M. 2008. Fate of sulfadiazine administered to pigs and its quantitative effect on the dynamics of bacterial resistance genes in manure and manured soil. *Soil Biology* and *Biochemistry*, 40(7): 1892–1900. https://doi.org/10.1016/j.soilbio.2008.03.014.

Hilscherova, K., Dusek, L., Kubik, V., Cupr, P., Hofman, J., Klanova, J. & Holoubek, I. 2007. Redistribution of organic pollutants in river sediments and alluvial soils related to major floods. *Journal of Soils* and *Sediments*, 7(3): 167–177. https://doi.org/10.1065/jss2007.04.222.

Holman, D. B., Hao, X., Topp, E., Yang, H. E. & Alexander, T. W. 2016. Effect of Co-Composting Cattle Manure with Construction and Demolition Waste on the Archaeal, Bacterial, and Fungal Microbiota, and on Antimicrobial Resistance Determinants. *PLOS ONE*, 11(6): e0157539. https://doi.org/10.1371/journal.pone.0157539.

Hölzel, C. S., Müller, C., Harms, K. S., Mikolajewski, S., Schäfer, S., Schwaiger, K. & Bauer, J. 2012. Heavy metals in liquid pig manure in light of bacterial antimicrobial resistance. *Environmental Research*, 113: 21–27. https://doi.org/10.1016/j.envres.2012.01.002.

Hoornweg, D. &Bhada-Tata, P. 2012. *Whatawaste. AGlobalreviewofsolidwaste management.* Knowledge papers. The World Bank. (also available at http: // documents. worldbank. org/curated/en/302341468126264791/pdf/68135-REVISED- What-a-Waste-2012-Final-updated.pdf).

Hope, B. K. 2006. Anexaminationofecologicalriskassessmentandmanagement practices. *Environment International*, 32(8): 983–995. https: //doi. org/10. 1016/j. envint. 2006. 06. 005.

Hopwood, D. A. 2007. Howdoantibiotic-producingbacteriaensuretheirself-resistance before antibiotic biosynthesis incapacitates them? *Molecular Microbiology*, 63(4): 937–940. https: //doi. org/10. 1111/j. 1365-2958. 2006. 05584. x.

Horckmans, L., Swennen, R., Deckers, J. &Maquil, R. 2005. Localbackgroundconcentrations of trace elements in soils: a case study in the Grand DuchyofLuxembourg. *CATENA*, 59(3): 279–304. https: //doi. org/10. 1016/j. catena. 2004. 09. 004.

Hosford, M. 2008. *Human health toxicological assessment of contaminants in soil: using sciencetocreateabetterplace*. SciencereportNo. SC050021/SR2. Bristol, Environment Agency. 70pp.

Hossain, M. F., White, S. K., Elahi, S. F., Sultana, N., Choudhury, M. H. K., Alam, Q. K., Rother, J. A. &Gaunt, J. L. 2005. TheefficiencyofnitrogenfertiliserforriceinBangladeshifarmers' fields. *Field Crops Research*, 93(1): 94–107. https: //doi. org/10. 1016/j. fcr. 2004. 09. 017.

Hough, R. L. 2007. Soil and human health: an epidemiological review. *European Journal of Soil Science*, 58(5): 1200–1212. https: //doi. org/10. 1111/j. 1365-2389. 2007. 00922. x.

Howard, B. J., Beresford, N. A., Barnett, C. L. &Fesenko, S. 2009. Quantifyingthetransferof radionuclides to food products from domestic farm animals. *Journal of Environmental Radioactivity*, 100(9): 767–773. https: //doi. org/10. 1016/j. jenvrad. 2009. 03. 010.

Hu, P., Huang, J., Ouyang, Y., Wu, L., Song, J., Wang, S., Li, Z., Han, C., Zhou, L., Huang, Y., Luo, Y. &Christie, P. 2013. Watermanagementaffectsarsenicandcadmiumaccumulation in different rice cultivars. *Environmental geochemistry* and *health*, 35(6): 767–778. https: //doi. org/10. 1007/s10653-013-9533-z.

Hu, Y., Cheng, H. &Tao, S. 2016. TheChallengesandSolutionsforCadmium- contaminated Rice in China: A Critical Review. *Environment International*, 92–93: 515–532. https: //doi. org/10. 1016/j. envint. 2016. 04. 042.

Huelster, A., Mueller, J. F. &Marschner, H. 1994. Soil-PlantTransferofPolychlorinated Dibenzo-p-dioxins and Dibenzofurans to Vegetables of the Cucumber Family (Cucurbitaceae). *Environmental Science &Technology*, 28(6): 1110–1115. https: //doi. org/10. 1021/es00055a021.

IAEA. 1998. *GuidelinesforIntegratedRiskAssessmentandManagementinLargeIndustrial Areas*. Vienna, INTERNATIONAL ATOMIC ENERGY AGENCY. (also available at http: //www-pub. iaea. org/books/IAEABooks/5649/Guidelines-for-Integrated-Risk-Assessment-and-Management-in-Large-Industrial-Areas).

Ingham, S. C., Losinski, J. A., Andrews, M. P., Breuer, J. E., Breuer, J. R., Wood, T. M. &Wright, T. H. 2004. Escherichia coli Contamination of Vegetables Grown in Soils Fertilized with Noncomposted Bovine Manure: Garden-Scale Studies. *Applied* and *Environmental Microbiology*, 70(11): 6420–6427. https: //doi. org/10. 1128/AEM. 70. 11. 6420-6427. 2004.

ISO. 2013. ISO11074: 2015-Soilquality——Vocabulary. (alsoavailableathttps: //www. iso. org/standard/59259. html).

Itai, T., Otsuka, M., Asante, K. A., Muto, M., Opoku-Ankomah, Y., Ansa-Asare, O. D. & Tanabe, S. 2014. Variation and distribution of metals and metalloids in soil/ash mixtures from Agbogbloshiee-waste recycling site in Accra, Ghana. *The Science of the Total Environment*, 470–471: 707–716. https: //doi. org/10. 1016/j. scitotenv. 2013. 10. 037.

Jacobsen, C. S. &Hjelmsø, M. H. 2014. Agricultural soils, pesticides andmicrobialdiversity. *Current Opinion in Biotechnology*, 27: 15–20. https: //doi. org/10. 1016/j. copbio. 2013. 09. 003.

Jaishankar, M., Tseten, T., Anbalagan, N., Mathew, B. B. &Beeregowda, K. N. 2014. Toxicity, mechanism and health effects of some heavy metals. *Interdisciplinary Toxicology*, 7(2). https: //doi. org/10. 2478/intox-2014-0009.

Jechalke, S., Heuer, H., Siemens, J., Amelung, W. &Smalla, K. 2014. Fateandeffectsof veterinary antibiotics in soil. *Trends in Microbiology*, 22(9): 536–545. https: //doi. org/10. 1016/j. tim. 2014. 05. 005.

JefaturadelEstado. 2001. PlanHidrológicoNacional. [Cited3April2018]. https: // www. boe. es/buscar/doc. php?id=BOE-A-2001-13042.

Jennings, A. A. 2013. Analysisofworldwideregulatoryguidancevaluesforthemost commonly regulated elemental surface soil contamination. *Journal of Environmental Management*, 118: 72–95. https: //doi. org/10. 1016/j. jenvman. 2012. 12. 032.

Jones, K. C. &deVoogt, P. 1999. Persistentorganicpollutants(POPs): stateofthe science. *EnvironmentalPollution(Barking, Essex: 1987)*, 100(1–3): 209–221.

Jones, O. A., Voulvoulis, N. &Lester, J. N. 2001. Humanpharmaceuticalsintheaquatic environment a review. *Environmental Technology*, 22(12): 1383–1394. https: //doi. org/10. 1080/09593332208618186.

Jordão, C. P., Nascentes, C. C., Cecon, P. R., Fontes, R. L. F. & Pereira, J. L. 2006. Heavy Metal Availability in Soil Amended with Composted Urban Solid Wastes. *Environmental Monitoring and Assessment*, 112(1–3): 309–326. https: //doi. org/10. 1007/s10661-006- 1072-y.

Joy, S. R., Bartelt-Hunt, S. L., Snow, D. D., Gilley, J. E., Woodbury, B. L., Parker, D. B., Marx, D. B. & Li, X. 2013. Fate and Transport of Antimicrobials and Antimicrobial Resistance Genes in

Soil and Runoff Following Land Application of Swine Manure Slurry. *Environmental Science &Technology*, 47(21): 12081–12088. https: //doi. org/10. 1021/ es4026358.

Juhasz, A. L., Smith, E., Weber, J., Rees, M., Rofe, A., Kuchel, T., Sansom, L. &Naidu, R. 2007. In vitro assessment of arsenic bioaccessibility in contaminated (anthropogenic and geogenic) soils. *Chemosphere*, 69(1): 69–78. https: //doi. org/10. 1016/j. chemosphere. 2007. 04. 046.

Kannan, K., Corsolini, S., Falandysz, J., Fillmann, G., Kumar, K. S., Loganathan, B. G., Mohd, M. A., Olivero, J., Wouwe, N. V., Yang, J. H. &Aldous, K. M. 2004. Perfluorooctanesulfonate and Related Fluorochemicals in Human Blood from Several Countries. *Environmental Science&Technology*, 38(17): 4489–4495. https: //doi. org/10. 1021/es0493446.

Kanter, D. R. 2018. Nitrogenpollution: akeybuildingblockforaddressingclimate change. *ClimaticChange*, 147(1–2): 11–21. https: //doi. org/10. 1007/s10584-017-2126-6.

Katz, D. 2016. UnderminingDemandManagementwithSupplyManagement: Moral HazardinIsraeliWaterPolicies. *Water*, 8(4): 159. https: //doi. org/10. 3390/w8040159.

Kemp, D. D. 1998. *The environment dictionary*. London；New York, Routledge.

Keraita, B. N. &Drechsel, P. 2004. Agriculturaluseofuntreatedurbanwastewater in Ghana. *In* C. A. Scott, N. I. Faruqui& L. Raschid-Sally, eds. *Wastewateruse in irrigated agriculture: confronting the livelihood* and *environmental realities*, pp. 101–112. Wallingford, CABI. (also available at http: //www. cabi. org/cabebooks/ebook/20043115023).

Keyte, I. J., Harrison, R. M. &Lammel, G. 2013. Chemical reactivity and long-range transport-potentialofpolyclicaromatichydrocarbons–areview. *ChemicalSociety Reviews*, 42(24): 9333. https: //doi. org/10. 1039/c3cs60147a.

Khachatourians, G. G. 1998. Agriculturaluseofantibioticsandtheevolutionand transfer of antibiotic-resistant bacteria. *CMAJ: Canadian Medical Associationjournal = journal de l'Associationmedicalecanadienne*, 159(9): 1129–1136.

Khalili, N. R., Scheff, P. A. &Holsen, T. M. 1995. PAHsourcefingerprintsforcokeovens, diesel and, gasoline engines, highway tunnels, and wood combustion emissions. *Atmospheric Environment*, 29(4): 533–542. https: //doi. org/10. 1016/1352-2310(94)00275-P.

Khan, A., Khan, S., Khan, M. A., Qamar, Z. &Waqas, M. 2015. Theuptakeandbioaccumulation of heavy metals by food plants, their effects on plants nutrients, and associated health risk: a review. *Environmental Science* and *Pollution Research*, 22(18): 13772–13799. https: //doi. org/10. 1007/s11356-015-4881-0.

Khan, S., Afzal, M., Iqbal, S. &Khan, Q. M. 2013. Plant–bacteriapartnershipsforthe remediation of hydrocarbon contaminated soils. *Chemosphere*, 90(4): 1317–1332. https: //doi. org/10. 1016/j. chemosphere. 2012. 09. 045.

Khandaghi, J., Razavilar, V. &Barzgari, A. 2010. IsolationofEscherichiacoliO157: H7 from manure fertilized farms and raw vegetables grown on it, in Tabriz city in Iran. *Afr. J. Microbiol. Res.*: 5.

Kim, E. J., Choi, S. -D. & Chang, Y. -S. 2011. Levels and patterns of polycyclicaromatichydrocarbons (PAHs) in soils after forest fires in South Korea. *Environmental Science andPollutionResearch*, 18(9): 1508–1517. https: //doi. org/10. 1007/s11356-011-0515-3.

Kim, H. S., Kim, K. -R., Kim, W. -I., Owens, G. & Kim, K. -H. 2017. InfluenceofRoadProximity on the Concentrations of Heavy Metals in Korean Urban Agricultural Soils and Crops. *Archives of Environmental Contamination* and *Toxicology*, 72(2): 260–268. https: // doi. org/10. 1007/ s00244-016-0344-y.

Kim, K. -H., Kabir, E. & Jahan, S. A. 2017. Exposuretopesticidesandtheassociated human health effects. *Science of The Total Environment*, 575: 525–535. https: //doi. org/10. 1016/j. scitotenv. 2016. 09. 009.

Kim, K. -R., Owens, G., Kwon, S. -I., So, K. -H., Lee, D. -B. & Ok, Y. S. 2011. Occurrenceand Environmental Fate of Veterinary Antibiotics in the Terrestrial Environment. *Water, Air, &SoilPollution*, 214(1–4): 163–174. https: //doi. org/10. 1007/s11270-010-0412-2.

Knapp, C. W., Callan, A. C., Aitken, B., Shearn, R., Koenders, A. & Hinwood, A. 2017. Relationship between antibiotic resistance genes and metals in residential soil samples from Western Australia. *Environmental Science* and *Pollution Research*, 24(3): 2484–2494. https: //doi. org/10. 1007/s11356-016-7997-y.

Knox, A., Seaman, J., Mench, M. & Vangronsveld, J. 2001. RemediationofMetal- and Radionuclides-Contaminated Soils by In Situ Stabilization Techniques. *Environmental Restoration of Metals-Contaminated Soils*, pp. 21–60. Taylor & Francis Group. (also available at https: //www. taylorfrancis. com/books/9781420026269/ chapters/10. 1201% 2F9781420026269-2).

Kobayashi, A., ed. 2012. *Geographiesofpeaceandarmedconflict*. London, Routledge. 241 pp.

Komárek, M., Cadková, E., Chrastný, V., Bordas, F. & Bollinger, J. -C. 2010. Contaminationof vineyard soils with fungicides: a review of environmental and toxicological aspects. *EnvironmentInternational*, 36(1): 138–151. https: //doi. org/10. 1016/j. envint. 2009. 10. 005.

Komárek, M., Vaněk, A. & Ettler, V. 2013. Chemicalstabilizationofmetalsandarsenicin contaminated soils using oxides——a review. *Environmental Pollution (Barking, Essex: 1987)*, 172: 9–22. https: //doi. org/10. 1016/j. envpol. 2012. 07. 045.

Komprda, J., Komprdová, K., Sáňka, M., Možný, M. & Nizzetto, L. 2013. InfluenceofClimate and Land Use Change on Spatially Resolved Volatilization of Persistent Organic Pollutants (POPs) from Background Soils. *Environmental Science & Technology*, 47(13): 7052–7059. https: //doi. org/10. 1021/es3048784.

Krapp, A. 2015. Plantnitrogenassimilationanditsregulation: acomplexpuzzlewith missing pieces. *Current Opinion in Plant Biology*, 25: 115–122. https: //doi. org/10. 1016/j. pbi. 2015. 05. 010.

Kukučka, P., Klánová, J., Sáňka, M. & Holoubek, I. 2009. Soilburdensofpersistentorganic pollutants – Their levels, fate and risk. Part II. Are there any trends in PCDD/F levels in mountain soils? *Environmental Pollution*, 157(12): 3255–3263. https://doi.org/10.1016/j.envpol.2009.05.029.

Kumar, A. & Maiti, S. K. 2015. Assessmentofpotentiallytoxicheavymetalcontamination in agricultural fields, sediment, and water from an abandoned chromite-asbestos mine waste of Roro hill, Chaibasa, India. *Environmental Earth Sciences*, 74(3): 2617– 2633. https://doi.org/10.1007/s12665-015-4282-1.

Kumar, K., Gupta, S. C., Baidoo, S. K., Chander, Y. & Rosen, C. J. 2005. Antibiotic Uptake by Plants from Soil Fertilized with Animal Manure. *Journal of Environment Quality*, 34(6): 2082. https://doi.org/10.2134/jeq2005.0026.

Kumar, V. & Kothiyal, N. C. 2016. AnalysisofPolycyclicAromaticHydrocarbon, Toxic Equivalency Factor and Related Carcinogenic Potencies in Roadside Soil within a Developing City of Northern India. *Polycyclic Aromatic Compounds*, 36(4): 506–526. https://doi.org/10.1080/10406638.2015.1026999.

Kumpiene, J., Lagerkvist, A. & Maurice, C. 2008. StabilizationofAs, Cr, Cu, PbandZnin soil using amendments--a review. *Waste Management (New York, N. Y.)*, 28(1): 215–225. https://doi.org/10.1016/j.wasman.2006.12.012.

Kuo, S., Ortiz-escobar, M. E., Hue, N. V. & Hummel, R. L. 2004. Composting and Compost Utilization for Agronomic and Container Crops. *Recent Developments in Environmental Biology*, 1: 451–513.

Kuppusamy, S., Kakarla, D., Venkateswarlu, K., Megharaj, M., Yoon, Y. -E. & Lee, Y. B. 2018. Veterinary antibiotics(VAs) contamination as a global agro-ecological issue: A critical view. *Agriculture, Ecosystems & Environment*, 257: 47–59. https://doi.org/10.1016/j.agee.2018.01.026.

Kuppusamy, S., Palanisami, T., Megharaj, M., Venkateswarlu, K. & Naidu, R. 2016. In- Situ Remediation Approaches for the Management of Contaminated Sites: A ComprehensiveOverview. *Reviews of Environmental Contamination and Toxicology*, 236: 1–115. https://doi.org/10.1007/978-3-319-20013-2_1.

Kuppusamy, S., Thavamani, P., Venkateswarlu, K., Lee, Y. B., Naidu, R. & Megharaj, M. 2017. Remediationapproachesforpolycyclicaromatichydrocarbons(PAHs)contaminated soils: Technological constraints, emerging trends and future directions. *Chemosphere*, 168: 944–968. https://doi.org/10.1016/j.chemosphere.2016.10.115.

terLaak, T. L., Agbo, S. O., Barendregt, A. & Hermens, J. L. M. 2006. FreelyDissolvedConcentrations of PAHs in SoilPore Water: Measurements via Solid-Phase Extraction and Consequencesfor

Soil Tests. *Environmental Science & Technology*, 40(4): 1307–1313. https: //doi. org/10. 1021/es0514803.

Lammoglia, S. -K., Kennedy, M. C., Barriuso, E., Alletto, L., Justes, E., Munier-Jolain, N. & Mamy, L. 2017. Assessing human health risks from pesticide use in convention al and innovative cropping systems with the BROWSE model. *Environment International*, 105: 66–78. https: //doi. org/10. 1016/j. envint. 2017. 04. 012.

Landrigan, P. J., Fuller, R., Acosta, N. J. R., Adeyi, O., Arnold, R., Basu, N. N., Baldé, A. B., Bertollini, R., Bose-O'Reilly, S., Boufford, J. I., Breysse, P. N., Chiles, T., Mahidol, C., Coll-Seck, A. M., Cropper, M. L., Fobil, J., Fuster, V., Greenstone, M., Haines, A., Hanrahan, D., Hunter, D., Khare, M., Krupnick, A., Lanphear, B., Lohani, B., Martin, K., Mathiasen, K. V., McTeer, M. A., Murray, C. J. L., Ndahimananjara, J. D., Perera, F., Potočnik, J., Preker, A. S., Ramesh, J., Rockström, J., Salinas, C., Samson, L. D., Sandilya, K., Sly, P. D., Smith, K. R., Steiner, A., Stewart, R. B., Suk, W. A., vanSchayck, O. C. P., Yadama, G. N., Yumkella, K. & Zhong, M. 2018. The Lancet Commission on pollution and health. *Lancet (London, England)*, 391(10119): 462–512. https: //doi. org/10. 1016/S0140-6736(17)32345-0.

Lauer, N. E., Harkness, J. S. & Vengosh, A. 2016. Brine Spills Associated with Unconventional Oil Development in North Dakota. *Environmental Science & Technology*, 50(10): 5389–5397. https: //doi. org/10. 1021/acs. est. 5b06349.

Lee, R. J., Strohmeier, B. R., Bunker, K. L. & VanOrden, D. R. 2008. Naturally occurring asbestos—A recurring public policy challenge. *Journal of Hazardous Materials*, 153(1–2): 1–21. https: //doi. org/10. 1016/j. jhazmat. 2007. 11. 079.

Lerda, D. 2011. Polycyclic Aromatic Hydrocarbons (PAHs) Factsheet., p. 34. Belgium, Joint Research Centre, European Commission. (also available at https: //ec. europa. eu/jrc/sites/jrcsh/files/Factsheet% 20PAH_0. pdf).

Lewis, S. E., Silburn, D. M., Kookana, R. S. & Shaw, M. 2016. Pesticide Behavior, Fate, and Effects in the Tropics: An Overview of the Current State of Knowledge. *Journal of Agricultural and Food Chemistry*, 64(20): 3917–3924. https: //doi. org/10. 1021/acs. jafc. 6b01320.

Li, A. 2009. PAHs in Comets: An Overview. In H. U. Käufl & C. Sterken, eds. *Deep Impact as a World Observatory Event: Synergies in Space, Time, and Wavelength*, pp. 161–175. Berlin, Heidelberg, Springer Berlin Heidelberg. (also available at http: //link. springer. com/10. 1007/978-3-540-76959-0_21).

Li, J. -S., Beiyuan, J., Tsang, D. C. W., Wang, L., Poon, C. S., Li, X. -D. & Fendorf, S. 2017. Arsenic- containing soil from geogenic source in Hong Kong: Leaching characteristics and stabilization/solidification. *Chemosphere*, 182: 31–39. https: //doi. org/10. 1016/j. chemosphere. 2017. 05. 019.

Li, Z., Ma, Z., vanderKuijp, T. J., Yuan, Z. &Huang, L. 2014. A review of soil heavy metal pollution from mines in China: Pollution and health risk assessment. *Science of The TotalEnvironment*, 468–469: 843–853. https: //doi. org/10. 1016/j. scitotenv. 2013. 08. 090.

Liang, Y., Bradford, S. A., Simunek, J., Heggen, M., Vereecken, H. &Klumpp, E. 2013. Retention and Remobilization of Stabilized Silver Nanoparticles in an Undisturbed Loamy Sand Soil. *Environmental Science &Technology*, 47(21): 12229–12237. https: //doi. org/10. 1021/es402046u.

Lin, C., Liu, J., Wang, R., Wang, Y., Huang, B. &Pan, X. 2013. PolycyclicAromatic Hydrocarbons in Surface Soils of Kunming, China: Concentrations, Distribution, Sources, and Potential Risk. *Soil* and *Sediment Contamination: An International Journal*, 22(7): 753–766. https: //doi. org/10. 1080/15320383. 2013. 768201.

Lindstrom, A. B., Strynar, M. J. &Libelo, E. L. 2011. PolyfluorinatedCompounds: Past, Present, and Future. *Environmental Science &Technology*, 45(19): 7954–7961. https: // doi. org/10. 1021/es2011622.

Lithner, D., Larsson, A. &Dave, G. 2011. Environmental and health hazard ranking and assessment of plastic polymers based on chemical composition. *The Science of the Total Environment*, 409(18): 3309–3324. https: //doi. org/10. 1016/j. scitotenv. 2011. 04. 038.

Liu, X., Zhang, W., Hu, Y., Hu, E., Xie, X., Wang, L. &Cheng, H. 2015. Arsenic pollution of agricultural soils by concentrated animal feeding operations (CAFOs). *Chemosphere*, 119: 273–281. https: //doi. org/10. 1016/j. chemosphere. 2014. 06. 067.

Logan, T. J. 2000. SoilsandEnvironmentalQuality. *In*M. E. Sumner, ed. *Handbookof SoilScience*, pp. G155–G170. BocaRaton, Fla, CRCPress.

Loganathan, B. G. &Lam, P. K. S., eds. 2012. *Globalcontaminationtrendsofpersistentorganic chemicals*. BocaRaton, CRCPress. 638pp.

Lu, C. &Tian, H. 2017. Globalnitrogenandphosphorusfertilizeruseforagriculture production in the past half century: shifted hot spots and nutrient imbalance. *Earth SystemScienceData*, 9(1): 181–192. https: //doi. org/10. 5194/essd-9-181-2017.

Lu, Y., Song, S., Wang, R., Liu, Z., Meng, J., Sweetman, A. J., Jenkins, A., Ferrier, R. C., Li, H., Luo, W. &Wang, T. 2015. Impactsofsoilandwaterpollutiononfoodsafetyand health risks in China. *Environment International*, 77: 5–15. https: //doi. org/10. 1016/j. envint. 2014. 12. 010.

Lucas, R. W., Klaminder, J., Futter, M. N., Bishop, K. H., Egnell, G., Laudon, H. &Högberg, P. 2011. A meta-analysis of the effects of nitrogen additions on base cations : Implications for plants, soils, and streams. *Forest Ecology* and *Management*, 262(2): 95–104.

Luo, L., Meng, H., Wu, R. &Gu, J. -D. 2017. Impactofnitrogenpollution/deposition on extracellular enzyme activity, microbial abundance and carbon storage in coastal mangrove sediment.

Chemosphere, 177: 275–283. https://doi.org/10.1016/j.chemosphere.2017.03.027.

Luo, Y., Wu, L., Liu, L., Han, C. & Li, Z. 2009. Heavy Metal Contamination and Remediation in Asian Agricultural Land. p. 9. Paper presented at MARCO Symposium, 2009, Japan.

Luque, J. 2014. Guía para la elaboración de los Planes de Descontaminación de Suelos.

Lv, B., Xing, M. & Yang, J. 2016. Speciation and transformation of heavy metals during vermicomposting of animal manure. *Bioresource Technology*, 209: 397–401. https://doi.org/10.1016/j.biortech.2016.03.015.

Lynn, M. 2017. Ways to Prevent Soil Pollution. In: *LIVESTRONG.COM* [online]. [Cited 3 April 2018]. https://www.livestrong.com/article/171421-ways-to-prevent-soil-pollution/.

Mackay, A. K., Taylor, M. P., Munksgaard, N. C., Hudson-Edwards, K. A. & Burn-Nunes, L. 2013. Identification of environmental lead sources and pathways in a mining and smelting town: Mount Isa, Australia. *Environmental Pollution*, 180: 304–311. https://doi.org/10.1016/j.envpol.2013.05.007.

Mansouri, A., Cregut, M., Abbes, C., Durand, M.-J., Landoulsi, A. & Thouand, G. 2017. The Environmental Issues of DDT Pollution and Bioremediation: a Multidisciplinary Review. *Applied Biochemistry* and *Biotechnology*, 181(1): 309–339. https://doi.org/10.1007/s12010-016-2214-5.

Mato, Y., Isobe, T., Takada, H., Kanehiro, H., Ohtake, C. & Kaminuma, T. 2001. Plastic resin pellets as a transport medium for toxic chemicals in the marine environment. *Environmental Science & Technology*, 35(2): 318–324.

McBratney, A., Field, D. J. & Koch, A. 2014. The dimensions of soil security. *Geoderma*, 213: 203–213. https://doi.org/10.1016/j.geoderma.2013.08.013.

McBride, M. B. 1994. *Environmental chemistry of soils*. New York, Oxford University Press.

McLaughlin, M. J., Palmer, L. T., Tiller, K. G., Beech, T. A. & Smart, M. K. 1994. Increased Soil Salinity Causes Elevated Cadmium Concentrations in Field-Grown Potato Tubers. *Journal of Environmental Quality*, 23(5): 1013–1018. https://doi.org/10.2134/jeq1994.00472425002300050023x.

McLaughlin, M. J., Parker, D. R. & Clarke, J. M. 1999. Metals and micronutrients–food safety issues. *Field Crops Research*, 60(1): 143–163. https://doi.org/10.1016/S0378-4290(98)00137-3.

McManus, P. S., Stockwell, V. O., Sundin, G. W. & Jones, A. L. 2002. Antibiotic use in plant agriculture. *Annual Review of Phytopathology*, 40(1): 443–465. https://doi.org/10.1146/annurev.phyto.40.120301.093927.

Meek, M., Boobis, A., Crofton, K., Heinemeyer, G., Raaij, M. & Vickers, C. 2011. Risk assessment of combined exposure to multiple chemicals: A WHO/IPCS framework. *Regulatory Toxicology* and *Pharmacology*, 60(2): S1–S14. https: //doi. org/10. 1016/j. yrtph. 2011. 03. 010.

Meharg, A. A. 2004. Arsenicinrice——understandinganewdisasterforSouth-EastAsia. *Trends in Plant Science*, 9(9): 415–417. https: //doi. org/10. 1016/j. tplants. 2004. 07. 002.

Mehra, R., Kumar, S., Sonkawade, R., Singh, N. P. &Badhan, K. 2010. Analysisofterrestrial naturally occurring radionuclides in soil samples from some areas of Sirsa district of Haryana, India using gamma ray spectrometry. *Environmental Earth Sciences*, 59(5): 1159–1164. https: //doi. org/10. 1007/s12665-009-0108-3.

Michael, I., Rizzo, L., McArdell, C. S., Manaia, C. M., Merlin, C., Schwartz, T., Dagot, C. &Fatta-Kassinos, D. 2013. Urbanwastewatertreatmentplantsashotspotsforthereleaseof antibiotics in the environment: A review. *Water Research*, 47(3): 957–995. https: //doi. org/10. 1016/j. watres. 2012. 11. 027.

Middeldorp, P. J. M., van Doesburg, W., Schraa, G. &Stams, A. J. M. 2005. Reductive dechlorination of hexachlorocyclohexane (HCH) isomers in soil under anaerobic conditions. *Biodegradation*, 16(3): 283–290. https: //doi. org/10. 1007/s10532-004-1573-8.

Miège, C., Choubert, J. M., Ribeiro, L., Eusèbe, M. &Coquery, M. 2009. Fateofpharmaceuticals and personal care products in wastewater treatment plants – Conception of a database and first results. *Environmental Pollution*, 157(5): 1721–1726. https: //doi. org/10. 1016/j. envpol. 2008. 11. 045.

Mielke, H. W. &Reagan, P. L. 1998. Soil is an important pathway of human lead exposure. *Environmental Health Perspectives*, 106 Suppl 1: 217–229.

Mileusnić, M., Mapani, B. S., Kamona, A. F., Ružičić, S., Mapaure, I. &Chimwamurombe, P. M. 2014. Assessment of agricultural soil contamination by potentially toxic metals dispersed from improperly disposed tailings, Kombat mine, Namibia. *Journal of GeochemicalExploration*, 144: 409–420. https: //doi. org/10. 1016/j. gexplo. 2014. 01. 009.

MINAM. 2017. ApruebanCriteriosparalaGestióndeSitiosContaminados-DECRETO SUPREMO-N ° 012-2017-MINAM. [Cited3April2018]. http: //busquedas. elperuano. pe/normaslegales/aprueban-criterios-para-la-gestion-de-sitios-contaminados-decreto-supremo-n-012-2017-minam-1593392-6/.

Minh, N. H., Minh, T. B., Kajiwara, N., Kunisue, T., Subramanian, A., Iwata, H., Tana, T. S., Baburajendran, R., Karuppiah, S., Viet, P. H., Tuyen, B. C. &Tanabe, S. 2006. Contamination by Persistent Organic Pollutants in Dumping Sites of Asian Developing Countries: Implication-ofEmergingPollutionSources. *ArchivesofEnvironmentalContamination and Toxicology*, 50(4): 474–481. https: //doi. org/10. 1007/s00244-005-1087-3.

Mirsal, I. 2008. *Soil Pollution: Origin, Monitoring & Remediation*. Springer Science & Business Media. 310 pp.

MMA. 2013. GuíaMetodológicaparalaGestióndeSuelosconPotencialPresencia de Contaminantes. http: //portal. mma. gob. cl/transparencia/mma/doc/Res_406_ GuiaMetodologicaSuelosContaminantes. pdf.

Morgan, R. 2013. Soil, heavymetals, andhumanhealth. *Soilsandhumanhealth*, pp. 59–82. BocaRaton, Fla, CRCPress.

Mortvedt, J. J. 1994. Plantandsoilrelationshipsofuraniumandthoriumdecayseries radionuclides-areview. *JournalofEnvironmentalQuality*, 23(4): 643–650.

Muir, D. C. G. &deWit, C. A. 2010. Trendsoflegacyandnewpersistentorganicpollutantsin the circumpolar arctic: Overview, conclusions, and recommendations. *Science of The Total Environment*, 408(15): 3044–3051. https: //doi. org/10. 1016/j. scitotenv. 2009. 11. 032.

Muñiz, O. 2008. *Los microelementosen la agricultura*. La Habana, Cuba, Agroinfo.

Muñoz, B. &Albores, A. 2011. DNA Damage Caused by PolycyclicAromatic Hydrocarbons: Mechanisms and Markers. *Selected Topics in DNA Repair*: 125–144.

Murakami, M., Nakagawa, F., Ae, N., Ito, M. &Arao, T. 2009. PhytoextractionbyRice Capable of Accumulating Cd at High Levels: Reduction of Cd Content of Rice Grain. *EnvironmentalScience&Technology*, 43(15): 5878–5883. https: //doi. org/10. 1021/ es8036687.

Mwakalapa, E. B., Mmochi, A. J., Müller, M. H. B., Mdegela, R. H., Lyche, J. L. &Polder, A. 2018. Occurrence and levels of persistent organic pollutants (POPs) in farmed and wild marine fish from Tanzania. A pilot study. *Chemosphere*, 191: 438–449. https: //doi. org/10. 1016/j. chemosphere. 2017. 09. 121.

Naidu, R., Channey, R., McConnell, S., Johnston, N., Semple, K. T., McGrath, S., Dries, V., Nathanail, P., Harmsen, J., Pruszinski, A., MacMillan, J. &Palanisami, T. 2015. Towardsbioavailability- based soil criteria: past, present and future perspectives. *Environmental Science* and *PollutionResearch*, 22(12): 8779–8785. https: //doi. org/10. 1007/s11356-013-1617-x.

Najeeb, U., Ahmad, W., Zia, M. H., Zaffar, M. &Zhou, W. 2017. Enhancingthelead phytostabilization in wetland plant Juncus effusus L. through somaclonal manipulation and EDTA enrichment. *Arabian Journal of Chemistry*, 10: S3310–S3317. https: //doi. org/10. 1016/j. arabjc. 2014. 01. 009.

Nathanail, P. 2011. Sustainableremediation: Quovadis?*RemediationJournal*, 21(4): 35–44. https: //doi. org/10. 1002/rem. 20298.

Navarro, I., delaTorre, A., Sanz, P., Porcel, M. Á., Pro, J., Carbonell, G. &Martínez, M. de L. Á. 2017. Uptake of perfluoroalkyl substances and halogenated flame retardants by crop plants grown in biosolids-amended soils. *Environmental Research*, 152: 199–206. https: //doi. org/10. 1016/j. envres. 2016. 10. 018.

Navarro, S., Vela, N. &Navarro, G. 2007. Review. Anoverviewontheenvironmental behaviour of pesticide residues in soils. *Spanish Journal of Agricultural Research*, 5(3): 357. https: //doi. org/10. 5424/sjar/2007053-5344.

Navas, A., Soto, J. &Machín, J. 2002. 238U, 226Ra, 210Pb, 232Thand40Kactivitiesin soil profiles of the Flysch sector (Central Spanish Pyrenees). *Applied Radiation* and *Isotopes*, 57(4): 579–589. https: //doi. org/10. 1016/S0969-8043(02)00131-8.

Nel, A., Xia, T., Madler, L. &Li, N. 2006. ToxicPotentialofMaterialsattheNanolevel. *Science*, 311(5761): 622–627. https: //doi. org/10. 1126/science. 1114397.

Nguyen, D. B., Rose, M. T., Rose, T. J., Morris, S. G. &vanZwieten, L. 2016. Impactof glyphosate on soil microbial biomass and respiration: A meta-analysis. *Soil Biology andBiochemistry*, 92: 50–57. https: //doi. org/10. 1016/j. soilbio. 2015. 09. 014.

Nicholson, F.., Chambers, B. ., Williams, J. . &Unwin, R.. 1999. Heavymetalcontentsof livestock feeds and animal manures in England and Wales. *Bioresource Technology*, 70(1): 23–31. https: //doi. org/10. 1016/S0960-8524(99)00017-6.

Nicholson, F. A., Smith, S. R., Alloway, B. J., Carlton-Smith, C. & Chambers, B. J. 2003. An inventory of heavy metals inputs to agricultural soils in England and Wales. *The Science of the Total Environment*, 311(1–3): 205–219. https: //doi. org/10. 1016/S0048- 9697(03)00139-6.

NICNAS. 1989. IndustrialChemicals(NotificationandAssessment)Act1989. [Cited3 April 2018]. https: //www. legislation. gov. au/Details/C2013C00643.

Nicolopoulou-Stamati, P., Maipas, S., Kotampasi, C., Stamatis, P. &Hens, L. 2016. Chemical Pesticides and Human Health: The Urgent Need for a New Concept in Agriculture. *Frontiers in Public Health*, 4. https: //doi. org/10. 3389/fpubh. 2016. 00148.

Nihei, N. 2013. Chapter 8. Radioactivity in Agricultural Products in Fukushima. *In* T. M. Nakanishi & K. Tanoi, eds. *Agricultural Implications of the Fukushima Nuclear Accident*, pp. 73–85. Tokyo, Springer Japan. (also available at http: //link. springer. com/10. 1007/978-4-431-54328-2).

Nisbet, A. F., Konoplev, A. V., Shaw, G., Lembrechts, J. F., Merckx, R., Smolders, E., Vandecasteele, C. M., Lönsjö, H., Carini, F. &Burton, O. 1993. Applicationoffertilisersandameliorantsto reduce soil to plant transfer of radiocaesium and radiostrontium in the medium to long term—a summary. *Science of The Total Environment*, 137(1): 173–182. https: //doi. org/10. 1016/0048-9697(93)90386-K.

Norton, G. J., Islam, M. R., Deacon, C. M., Zhao, F. -J., Stroud, J. L., McGrath, S. P., Islam, S., Jahiruddin, M., Feldmann, J., Price, A. H. &Meharg, A. A. 2009. Identification oflowinorganic and total grain arsenic rice cultivars from Bangladesh. *Environmental Science &Technology*, 43(15): 6070–6075.

Ockleford, C., Adriaanse, P., Berny, P., Brock, T., Duquesne, S., Grilli, S., HernandezJerez, A. F., Bennekou, S. H., Klein, M., Kuhl, T., Laskowski, R., Machera, K., Pelkonen, O., Pieper, S., Stemmer, M., Sundh, I., Teodorovic, I., Tiktak, A., Topping, C. J., Wolterink, G., Craig, P., deJong, F., Manachini, B., Sousa, P., Swarowsky, K., Auteri, D., Arena, M. &Rob, S. 2017. Scientific Opinion addressing the state of the science on risk assessment of plant protection products for in-soil organisms. *EFSA Journal*, 15(2). https: //doi. org/10. 2903/j. efsa. 2017. 4690.

Odabasi, M., Dumanoglu, Y., OzgunergeFalay, E., Tuna, G., Altiok, H., Kara, M., Bayram, A., Tolunay, D. &Elbir, T. 2016. Investigationofspatialdistributionsandsources of persistent organic pollutants (POPs) in a heavily polluted industrial region using tree components. *Chemosphere*, 160: 114–125. https: //doi. org/10. 1016/j. chemosphere. 2016. 06. 076.

Odongo, A. O., Moturi, W. N. &Mbuthia, E. K. 2016. Heavymetalsandparasiticgeohelminths toxicity among geophagous pregnant women: a case study of Nakuru Municipality, Kenya. *Environmental Geochemistry* and *Health*, 38(1): 123–131. https: //doi. org/10. 1007/ s10653-015-9690-3.

OECD. 2008. *OECD Environmental Outlook to 2030*. OECD Environmental Outlook. OECD Publishing. (also available at http: //www. oecd-ilibrary. org/environment/ oecd-environmental-outlook-to-2030_9789264040519-en).

Ogundele, L. T., Owoade, O. K., Hopke, P. K. &Olise, F. S. 2017. Heavymetalsinindustrially emitted particulate matter in Ile-Ife, Nigeria. *Environmental Research*, 156: 320–325. https: //doi. org/10. 1016/j. envres. 2017. 03. 051.

Okere, U. V. 2011. BiodegradationofPAHsinPristineSoilsfromDifferentClimatic Regions. *Journal of Bioremediation & Biodegradation*, s1. https: //doi. org/10. 4172/2155- 6199. S1-006.

Oldeman, L. R. 1991. Worldmaponstatusofhuman-inducedsoildegradation. Nairobi, Kenya : Wageningen, Netherlands, UNEP；ISRIC.

Oliver, D., Schultz, J., Tiller, K. &Merry, R. 1993. Theeffectofcroprotationsandtillage practices on cadmium concentration in wheat grain. *Australian Journal of Agricultural Research*, 44(6): 1221. https: //doi. org/10. 1071/AR9931221.

Oliver, D. P., Hannam, R., Tiller, K. G., Wilhelm, N. S., Merry, R. H. &Cozens, G. D. 1994. The Effects of Zinc Fertilization on Cadmium Concentration in Wheat Grain. *Journal of Environmental Quality*, 23(4): 705–711. https: //doi. org/10. 2134/ jeq1994. 00472425002300040013x.

Oliver, M. A. &Gregory, P. J. 2015. Soil, foodsecurityandhumanhealth: areview: Soil, food security and human health. *European Journal of Soil Science*, 66(2): 257–276. https: //doi. org/10. 1111/ ejss. 12216.

O'Neill, J. 2014. ReviewonAntimicrobialResistance: Tacklingacrisisforthehealth and wealth of nations. Review on Antimicrobial Resistance. London. (also available athttps: //amr-review. org/ Publications. html).

Ongeng, D., Vasquez, G. A., Muyanja, C., Ryckeboer, J., Geeraerd, A. H. &Springael, D. 2011. Transfer and internalisation of Escherichia coli O157: H7 and Salmonella enterica serovar Typhimurium in cabbage cultivated on contaminated manure-amended soil under tropical field conditions in Sub-Saharan Africa. *International JournalofFoodMicrobiology*, 145(1): 301–310. https: //doi. org/10. 1016/j. ijfoodmicro. 2011. 01. 018.

Pan, B. &Xing, B. 2008. AdsorptionMechanismsofOrganicChemicalsonCarbon Nanotubes. *Environmental Science &Technology*, 42(24): 9005–9013. https: //doi. org/10. 1021/es801777n.

Pan, B. &Xing, B. 2012. Applicationsandimplicationsofmanufacturednanoparticles insoils: areview. *EuropeanJournalofSoilScience*, 63(4): 437–456. https: //doi. org/10. 1111/ j. 1365-2389. 2012. 01475. x.

Pan, J., Plant, J. A., Voulvoulis, N., Oates, C. J. &Ihlenfeld, C. 2010. Cadmiumlevelsin Europe: implications for human health. *Environmental Geochemistry* and *Health*, 32(1): 1–12. https: // doi. org/10. 1007/s10653-009-9273-2.

Panagiotakis, I. &Dermatas, D. 2015. RemediationofContaminatedSites. *Bulletinof Environmental Contamination* and *Toxicology*, 94(3): 267–268. https: //doi. org/10. 1007/ s00128-015-1490-z.

Park, J. H., Lamb, D., Paneerselvam, P., Choppala, G., Bolan, N. & Chung, J. -W. 2011. Role of organic amendments on enhanced bioremediation of heavy metal(loid) contaminated soils. *Journal of Hazardous Materials*, 185(2–3): 549–574. https: //doi. org/10. 1016/j. jhazmat. 2010. 09. 082.

Passatore, L., Rossetti, S., Juwarkar, A. A. &Massacci, A. 2014. Phytoremediationand bioremediation of polychlorinated biphenyls (PCBs): state of knowledge and research perspectives. *Journal of Hazardous Materials*, 278: 189–202. https: //doi. org/10. 1016/j. jhazmat. 2014. 05. 051.

Paye, H. deS., Mello, J. W. V. de&Melo, S. B. de. 2012. Métodosdeanálisemultivariadano estabelecimento de valores de referênciadequalidade para elementos-traçoemsolos. *RevistaBrasileiradeCiênciadoSolo*, 36(3): 1031–1042. https: //doi. org/10. 1590/ S0100-06832012000300033.

Paz-Alberto, A. M. &Sigua, G. C. 2013. Phytoremediation: AGreenTechnologyto Remove Environmental Pollutants. *American Journal of Climate Change*, 02(01): 71– 86. https: //doi. org/10. 4236/ajcc. 2013. 21008.

Pedrero, F., Kalavrouziotis, I., Alarcón, J. J., Koukoulakis, P. &Asano, T. 2010. Useoftreated municipal wastewater in irrigated agriculture—Review of some practices in Spain andGreece. *AgriculturalWaterManagement*, 97(9): 1233–1241. https: //doi. org/10. 1016/j. agwat. 2010. 03. 003.

Perkins, A. E. & Nicholson, W. L. 2008. Uncovering New Metabolic Capabilities of Bacillus subtilis Using Phenotype Profiling of Rifampin-Resistant rpoB Mutants. *Journal of Bacteriology*, 190(3): 807–814. https: //doi. org/10. 1128/JB. 00901-07.

Perkins, D. N., BruneDrisse, M. -N., Nxele, T. &Sly, P. D. 2014. E-Waste: AGlobalHazard. *Annals of Global Health*, 80(4): 286–295. https: //doi. org/10. 1016/j. aogh. 2014. 10. 001.

Petrie, B., Barden, R. &Kasprzyk-Hordern, B. 2015. A reviewonemergingcontaminantsin wastewaters and the environment: Current knowledge, understudied areas and recommendations for future monitoring. *Water Research*, 72: 3–27. https: //doi. org/10. 1016/j. watres. 2014. 08. 053.

Pierzynski, G. M., Sims, J. T. &Vance, G. F. 2005. *Soilsandenvironmentalquality*. 2nded edition. BocaRaton, CRCPress. 459pp.

Pietrzak, U. &McPhail, D. C. 2004. Copperaccumulation, distributionandfractionation in vineyard soils of Victoria, Australia. *Geoderma*, 122(2–4): 151–166. https: //doi. org/10. 1016/j. geoderma. 2004. 01. 005.

du Plessis, E. M., Duvenage, F. & Korsten, L. 2015. Determining the Potential Link between Irrigation Water Quality and the Microbiological Quality of Onions by Phenotypic and Genotypic Characterization of Escherichia coli Isolates. *Journal of Food Protection*, 78(4): 643–651. https: //doi. org/10. 4315/0362-028X. JFP-14-486.

Podolský, F., Ettler, V., Šebek, O., Ježek, J., Mihaljevič, M., Kříbek, B., Sracek, O., Vaněk, A., Penížek, V., Majer, V., Mapani, B., Kamona, F. &Nyambe, I. 2015. Mercuryinsoilprofiles from metal mining and smelting areas in Namibia and Zambia: distribution and potentialsources. *JournalofSoilsandSediments*, 15(3): 648–658. https: //doi. org/10. 1007/ s11368-014-1035-9.

Popp, J., Pető, K. &Nagy, J. 2013. Pesticideproductivityandfoodsecurity. Areview. *Agronomy for Sustainable Development*, 33(1): 243–255. https: //doi. org/10. 1007/s13593- 012-0105-x.

Posthuma, L., Eijsackers, H. J. P., Koelmans, A. A. &Vijver, M. G. 2008. Ecologicaleffectsof diffuse mixed pollution are site-specific and require higher-tier risk assessment to improve site management decisions: A discussion paper. *Science of The Total Environment*, 406(3): 503–517. https: //doi. org/10. 1016/j. scitotenv. 2008. 06. 065.

Prestt, I., Jefferies, D. J. &Moore, N. W. 1970. Polychlorinatedbiphenylsinwildbirdsin Britain and their avian toxicity. *Environmental Pollution (1970)*, 1(1): 3–26. https: //doi. org/10. 1016/0013-9327(70)90003-0.

Pretty, J. N., Mason, C. F., Nedwell, D. B., Hine, R. E., Leaf, S. &Dils, R. 2003. Environmental Costs of Freshwater Eutrophication in England and Wales. *Environmental Science &Technology*, 37(2): 201–208. https: //doi. org/10. 1021/es020793k.

Provoost, J., Cornelis, C. &Swartjes, F. 2006. ComparisonofSoilClean-upStandards for Trace

Elements Between Countries: Why do they differ? (9 pages). *Journal of SoilsandSediments*, 6(3): 173–181. https: //doi. org/10. 1065/jss2006. 07. 169.

Puglisi, E. 2012. Response of microbial organisms (aquatic and terrestrial) to pesticides. *EFSA Supporting Publications*, 9(11). https: //doi. org/10. 2903/sp. efsa. 2012. EN-359.

Puschenreiter, M., Horak, O., Friesl, W. &Hartl, W. 2005. Low-costagriculturalmeasures to reduce heavy metal transfer into the food chain - a review. *Plant, Soil* and *Environment*, 51(1): 1–11.

Raffa, D., Tubiello, F., Turner, D. &MonteroSerrano, J. 2018. Nitrogeninputsto agricultural soils from livestock manure. New statistics., p. 86. Rome, Italy, Food and Agriculture Organization of the United Nations. (also available at http: //www. fao. org/3/I8153EN/i8153en. pdf).

Randhawa, M. A., Salim-ur-Rehman, Anjum, F. M. & Awan, J. A. 2014. Pesticide residuesinfood: Healthimplicationsforchildrenandwomen. *In*R. Bhat&V. M. Gomez-Lopez, eds. *Practical Food Safety: Contemporary Issues* and *Future Directions*, pp. 145–167. Hoboken, UNITED KINGDOM, John Wiley & Sons, Incorporated. (also available athttp: //ebookcentral. proquest. com/lib/bull-ebooks/detail. action?docID=1655929).

Rankin, K., Mabury, S. A., Jenkins, T. M. &Washington, J. W. 2016. ANorthAmericanand global survey of perfluoroalkyl substances in surface soils: Distribution patterns and mode of occurrence. *Chemosphere*, 161: 333–341. https: //doi. org/10. 1016/j. chemosphere. 2016. 06. 109.

Rao, G., Lu, C. &Su, F. 2007. Sorptionofdivalentmetalionsfromaqueoussolutionby carbon nanotubes: A review. *Separation* and *Purification Technology*, 58(1): 224–231. https: //doi. org/10. 1016/j. seppur. 2006. 12. 006.

Ratcliffe, D. A. 1970. ChangesAttributabletoPesticidesinEggBreakageFrequency andEggshellThicknessinSomeBritishBirds. *TheJournalofAppliedEcology*, 7(1): 67. https: //doi. org/10. 2307/2401613.

Reeuwijk, N. M., Talidda, A., Malisch, R., Kotz, A., Tritscher, A., Fiedler, H., Zeilmaker, M. J., Kooijman, M., Wienk, K. J. H., Traag, W. A. &Hoogenboom, R. L. A. P. 2013. Dioxins (polychlorinated dibenzo-p-dioxins and polychlorinated dibenzo-furans) in traditional clay products used during pregnancy. *Chemosphere*, 90(5): 1678–1685. https: //doi. org/10. 1016/j. chemosphere. 2012. 09. 064.

Reeves, W. R., Barhoumi, R., Burghardt, R. C., Lemke, S. L., Mayura, K., McDonald, T. J., Phillips, T. D. &Donnelly, K. C. 2001. EvaluationofMethodsforPredictingtheToxicityof Polycyclic Aromatic Hydrocarbon Mixtures. *Environmental Science &Technology*, 35(8): 1630–1636. https: //doi. org/10. 1021/es001689a.

Reimann, C., Filzmoser, P. & Garrett, R. G. 2005. Background and threshold: critical comparison of methods of determination. *Science of The Total Environment*, 346(1–3): 1–16. https: //doi.

org/10. 1016/j. scitotenv. 2004. 11. 023.

Ren, X., Chen, C., Nagatsu, M. &Wang, X. 2011. Carbonnanotubesasadsorbentsin environmental pollution management: A review. *Chemical Engineering Journal*, 170(2–3): 395–410. https: // doi. org/10. 1016/j. cej. 2010. 08. 045.

Rensing, C. &Pepper, I. L. 2006. Chapter 30. Antibiotic-Resistant Bacteria and Gene Transfer. *Environmental* and *Pollution Science*, pp. 499–505. San Diego, UNITED STATES, Elsevier Science &Technology. (also available at http: //ebookcentral. proquest. com/lib/bull-ebooks/detail. action?docID=297063).

Rigol, A., Vidal, M. &Rauret, G. 2002. Anoverviewoftheeffectoforganicmatteron soil–radiocaesium interaction: implications in root uptake. *Journal of Environmental Radioactivity*, 58(2–3): 191–216. https: //doi. org/10. 1016/S0265-931X(01)00066-2.

Rillig, M. C. 2012. MicroplasticinTerrestrialEcosystemsandtheSoil?*Environmental Science&Technology*, 46(12): 6453–6454. https: //doi. org/10. 1021/es302011r.

Rizwan, M., Ali, S., Adrees, M., Ibrahim, M., Tsang, D. C. W., Zia-ur-Rehman, M., Zahir, Z. A., Rinklebe, J., Tack, F. M. G. &Ok, Y. S. 2017. Acriticalreviewoneffects, tolerancemechanisms and management of cadmium in vegetables. *Chemosphere*, 182: 90–105. https: //doi. org/10. 1016/j. chemosphere. 2017. 05. 013.

Robinson, B. H. 2009. E-waste: Anassessmentofglobalproductionandenvironmental impacts. *ScienceofTheTotalEnvironment*, 408(2): 183–191. https: //doi. org/10. 1016/j. scitotenv. 2009. 09. 044.

Rocha-Santos, T. &Duarte, A. C. 2015. Acriticaloverviewoftheanalyticalapproaches to the occurrence, the fate and the behavior of microplastics in the environment. *TrACTrendsinAnalyticalChemistry*, 65: 47–53. https: //doi. org/10. 1016/j. trac. 2014. 10. 011.

Rodrigues, S. M., Pereira, M. E., Duarte, A. C. &Römkens, P. F. A. M. 2012. Soil–plant–animal transfer models to improve soil protection guidelines: A case study from Portugal. *EnvironmentInternational*, 39(1): 27–37. https: //doi. org/10. 1016/j. envint. 2011. 09. 005.

Romero-Freire, A., MartinPeinado, F. J. &vanGestel, C. A. M. 2015. Effectofsoilproperties on the toxicity of Pb: Assessment of the appropriateness of guideline values. *Journal ofHazardousMaterials*, 289: 46–53. https: //doi. org/10. 1016/j. jhazmat. 2015. 02. 034.

Rosal, R., Rodea-Palomares, I., Boltes, K., Fernández-Piñas, F., Leganés, F. &Petre, A. 2010. Ecotoxicological assessment of surfactants in the aquatic environment: Combined toxicity of docusate sodium with chlorinated pollutants. *Chemosphere*, 81(2): 288– 293. https: //doi. org/10. 1016/j. chemosphere. 2010. 05. 050.

Rosendahl, I., Siemens, J., Groeneweg, J., Linzbach, E., Laabs, V., Herrmann, C., Vereecken, H. &Amelung, W. 2011. Dissipation and Sequestration of theVeterinaryAntibiotic Sulfadiazine and Its Metabolites under Field Conditions. *Environmental Science &Technology*, 45(12): 5216–5222. https: //doi. org/10. 1021/es200326t.

Safronova, V. I., Stepanok, V. V., Engqvist, G. L., Alekseyev, Y. V. &Belimov, A. A. 2006. Root-associated bacteria containing 1-aminocyclopropane-1-carboxylate deaminase improve growth and nutrient uptake by pea genotypes cultivated in cadmium supplemented soil. *Biology and Fertility of Soils*, 42(3): 267–272. https: //doi. org/10. 1007/ s00374-005-0024-y.

Saha, J. K., Selladurai, R., Coumar, M. V., Dotaniya, M. L., Kundu, S. & Patra, A. K. 2017. *Soil Pollution - An Emerging Threat to Agriculture*. Environmental Chemistry fora Sustainable World. Singapore, Springer Singapore. (also available athttp: //link. springer. com/10. 1007/978-981-10-4274-4).

Salminen, R. &Gregorauskiene, V. 2000. Considerationsregardingthedefinitionof a geochemical baseline of elements in the surficial materials in areas differing in basic geology. *Applied Geochemistry*, 15(5): 647–653. https: //doi. org/10. 1016/S0883- 2927(99)00077-3.

Santos, A. &Flores, M. 1995. Effectsofglyphosateonnitrogenfixationoffree-living heterotrophic bacteria. *Letters in Applied Microbiology*, 20(6): 349–352. https: //doi. org/10. 1111/j. 1472-765X. 1995. tb01318. x.

Sarigiannis, D. A. &Hansen, U. 2012. Consideringthecumulativeriskofmixturesof chemicals – A challenge for policy makers. *Environmental Health*, 11(Suppl 1): S18. https: //doi. org/10. 1186/1476-069X-11-S1-S18.

Sarmah, A. K., Meyer, M. T. &Boxall, A. B. A. 2006. Aglobalperspectiveontheuse, sales, exposure pathways, occurrence, fate and effects of veterinary antibiotics(VAs) in the environment. *Chemosphere*, 65(5): 725–759. https: //doi. org/10. 1016/j. chemosphere. 2006. 03. 026.

Sassman, S. A. &Lee, L. S. 2005. SorptionofThreeTetracyclinesbySeveralSoils: Assessing the Role of pH and Cation Exchange. *Environmental Science &Technology*, 39(19): 7452–7459. https: // doi. org/10. 1021/es0480217.

Sauvé, S. & Desrosiers, M. 2014. A review of what is an emerging contaminant. *Chemistry Central Journal*, 8(1): 15. https: //doi. org/10. 1186/1752-153X-8-15.

Scallan, E., Griffin, P. M., Angulo, F. J., Tauxe, R. V. &Hoekstra, R. M. 2011. FoodborneIllness Acquired in the United States—Unspecified Agents. *Emerging Infectious Diseases*, 17(1): 16–22. https: //doi. org/10. 3201/eid1701. P21101.

Schafer, R. 1995. Results of the contaminated-site survey. pp. 309–322. Paper presented at Proceedings of the International Workshop on Military and Armament Contaminated Sites, 1995, Berlin.

Schmidt, C. 2010. How PCBs Are Like Grasshoppers. *Environmental Science & Technology*, 44(8): 2752–2752. https: //doi. org/10. 1021/es100696y.

Schnug, E. &Lottermoser, B. G. 2013. Fertilizer-Derived Uranium and its Threat to Human Health. *Environmental Science &Technology*, 47(6): 2433–2434. https: //doi. org/10. 1021/es4002357.

Science Communication Unit, University of the West of England. 2013. Science for Environment Policy In-depth Report: Soil Contamination: Impacts on Human Health. Bristol, UK, European Commission DG Environment. (also available at http: //ec. europa. eu/environment/integration/research/newsalert/pdf/IR5_en. pdf).

Scott, K.., Ashley, P.. &Lawie, D.. 2001. The geochemistry, mineralogy and maturity of gossans derived from volcanogenic Zn–Pb–Cu deposits of the eastern Lachlan Fold Belt, NSW, Australia. *Journal of Geochemical Exploration*, 72(3): 169–191. https: //doi. org/10. 1016/S0375-6742(01)00159-5.

Scullion, J. 2006. Remediating polluted soils. *Naturwissenschaften*, 93(2): 51–65. https: // doi. org/10. 1007/s00114-005-0079-5.

Shacklette, H. T. &Boerngen, J. G. 1984. Element Concentrations in Soils and Other Surficial Materials of the Conterminous United States., p. 63. US Geological Survey.

Shaw, G. 1993. Blockade by fertilisers of caesium and stronium uptake into crops: effects on the root uptake process. *Science of The Total Environment*, 137(1): 119–133. https: //doi. org/10. 1016/0048-9697(93)90381-F.

Shayler, H., McBride, M. &Harrison, E. 2009. Sources and Impacts of Contaminants in Soils. (also available at http: //ecommons. cornell. edu/handle/1813/14282).

Shen, W., Lin, X., Shi, W., Min, J., Gao, N., Zhang, H., Yin, R. &He, X. 2010. Higher rates of nitrogen fertilization decrease soil enzyme activities, microbial functional diversity and nitrification capacity in a Chinese polytunnel greenhouse vegetable land. *Plant andSoil*, 337(1–2): 137–150. https: //doi. org/10. 1007/s11104-010-0511-2.

Shiralipour, A., McConnell, D. B. &Smith, W. H. 1992. Uses and benefits of MSW compost: A review and an assessment. *Biomass* and *Bioenergy*, 3(3–4): 267–279. https: //doi. org/10. 1016/0961-9534(92)90031-K.

Simonich, S. L. &Hites, R. A. 1995. Organic Pollutant Accumulation in Vegetation. *Environmental Science & Technology*, 29(12): 2905–2914. https: //doi. org/10. 1021/ es00012a004.

Singh, D. K., ed. 2012. *Toxicology: Agriculture And Environment: Pesticide Chemistry And Toxicology*. BENTHAM SCIENCE PUBLISHERS. (also available at http: //www. eurekaselect. com/50654/volume/1).

Singh, R., Singh, A., Misra, V. &Singh, R. 2011. Degradation of Lindane Contaminated Soil Using Zero-Valent Iron Nanoparticles. *Journal of Biomedical Nanotechnology*, 7(1): 175–176. https: //doi. org/10. 1166/jbn. 2011. 1256.

Šmídová, K., Hofman, J., Ite, A. E. &Semple, K. T. 2012. Fate and bioavailability of 14Cpyrene and 14C-lindane in sterile natural and artificial soils and the influence of aging. *Environmental Pollution*, 171: 93–98. https: //doi. org/10. 1016/j. envpol. 2012. 07. 031.

Smit, E., Elsas, J. D. van, Veen, J. A. van & Vos, W. M. de. 1991. Detection of Plasmid Transfer from Pseudomonas fluorescens to Indigenous Bacteria in Soil by Using Bacteriophage φR2f for Donor Counterselection. *Applied* and *Environmental Microbiology*, 57(12): 3482–3488.

Smith, J. T., Comans, R. N. J., Beresford, N. A., Wright, S. M., Howard, B. J. & Camplin, W. C. 2000. Chernobyl's legacy in food and water: Pollution. *Nature*, 405(6783): 141–141. https: //doi. org/10. 1038/35012139.

Smith, S. 2009. A critical review of the bioavailability and impacts of heavy metals in municipal solid waste composts compared to sewage sludge. *Environment International*, 35(1): 142–156. https: //doi. org/10. 1016/j. envint. 2008. 06. 009.

Souza Arroyo, V., Martínez Flores, K., Bucio Ortiz, L., Gómez-Quiroz, L. E. & Gutiérrez-Ruiz, M. C. 2012. Liver and Cadmium Toxicity. *Journal of Drug Metabolism & Toxicology*, 03(06). https: //doi. org/10. 4172/2157-7609. S5-001.

Srogi, K. 2007. Monitoring of environmental exposure to polycyclic aromatic hydrocarbons: a review. *Environmental Chemistry Letters*, 5(4): 169–195. https: //doi. org/10. 1007/s10311-007-0095-0.

SSR. 2010. Soil Contamination in West Africa Environmental Remediation Pollution. (also available at https: //www. scribd. com/doc/71599035/Soil- Contamination-in-West-Africa).

Stasinos, S. & Zabetakis, I. 2013. The uptake of nickel and chromium from irrigation water by potatoes, carrots and onions. *Ecotoxicology* and *Environmental Safety*, 91: 122–128. https: //doi. org/10. 1016/j. ecoenv. 2013. 01. 023.

Steinnes, E., Allen, R. O., Petersen, H. M., Rambæk, J. P. & Varskog, P. 1997. Evidence of large scale heavy-metal contamination of natural surface soils in Norway from long- range atmospheric transport. *Science of The Total Environment*, 205(2–3): 255–266. https: //doi. org/10. 1016/S0048-9697(97)00209-X.

Steinnes, E., Berg, T. & Uggerud, H. T. 2011. Three decades of atmospheric metal deposition in Norway as evident from analysis of moss samples. *Science of The Total Environment*, 412–413: 351–358. https: //doi. org/10. 1016/j. scitotenv. 2011. 09. 086.

Stewart, W. M., Dibb, D. W., Johnston, A. E. & Smyth, T. J. 2005. The Contribution of Commercial Fertilizer Nutrients to Food Production. *Agronomy Journal*, 97(1): 1. https: //doi. org/10. 2134/agronj2005. 0001.

Sthiannopkao, S. & Wong, M. H. 2013. Handling e-waste in developed and developing countries: Initiatives, practices, and consequences. *Science of The Total Environment*, 463–464: 1147–1153. https: //doi. org/10. 1016/j. scitotenv. 2012. 06. 088.

Stockholm Convention. 2018. *All POPs listed in the Stockholm Convention.* [online]. [Cited 3 April 2018]. http: //chm. pops. int/TheConvention/ThePOPs/ListingofPOPs/ tabid/2509/Default. aspx.

Stork, P. R. &Lyons, D. J. 2012. Phosphoruslossandspeciationinoverlandflowfrom a plantation horticulture catchment and in an adjoining waterway in coastal Queensland, Australia. *SoilResearch*, 50(6): 515. https: //doi. org/10. 1071/SR12042.

Stratton, M. L., Barker, A. V. &Rechcigl, J. E. 1995. Compost. *In*J. E. Rechcigl, ed. *Soil amendments* and *environmental quality*, p. Agriculture &environment series. New York, LewisPublishers.

Štrok, M. &Smodiš, B. 2012. Transferofnaturalradionuclidesfromhayandsilage to cow's milk in the vicinity of a former uranium mine. *Journal of Environmental Radioactivity*, 110: 64–68. https: //doi. org/10. 1016/j. jenvrad. 2012. 02. 009.

Strzebońska, M., Jarosz-Krzemińska, E. &Adamiec, E. 2017. AssessingHistoricalMining and Smelting Effects on Heavy Metal Pollution of River Systems over Span of Two Decades. *Water, Air, &SoilPollution*, 228(4). https: //doi. org/10. 1007/s11270-017-3327-3.

Sukul, P., Lamshöft, M., Kusari, S., Zühlke, S. &Spiteller, M. 2009. Metabolismandexcretion kinetics of 14C-labeled and non-labeled difloxacin in pigs after oral administration, and antimicrobial activity of manure containing difloxacin and its metabolites. *Environmental Research*, 109(3): 225–231. https: //doi. org/10. 1016/j. envres. 2008. 12. 007.

Sun, F., Ma, Y., Guo, H. &Ji, R. 2018. FateofSeveralTypicalOrganicPollutantsinSoiland Impacts of Earthworms and Plants. *In* Y. Luo & C. Tu, eds. *Twenty Years of Research* and *Development on Soil Pollution* and *Remediation in China*, pp. 575–589. Singapore, Springer Singapore. (also available at http: //link. springer. com/10. 1007/978-981-10-6029-8_35).

Swartjes, F. A., ed. 2011. *Dealing with Contaminated Sites.* Dordrecht, Springer Netherlands. (alsoavailableathttp: //link. springer. com/10. 1007/978-90-481-9757-6).

Swartjes, F. A., Rutgers, M., Lijzen, J. P. A., Janssen, P. J. C. M., Otte, P. F., Wintersen, A., Brand, E. &Posthuma, L. 2012. StateoftheartofcontaminatedsitemanagementinThe Netherlands: Policy framework and risk assessment tools. *Science of The Total Environment*, 427–428: 1–10. https: //doi. org/10. 1016/j. scitotenv. 2012. 02. 078.

Swartjes, F. A. &Tromp, P. C. 2008. ATieredApproachfortheAssessmentoftheHuman Health Risks of Asbestos in Soils. *Soil* and *Sediment Contamination: An International Journal*, 17(2): 137–149. https: //doi. org/10. 1080/15320380701870484.

Swati, Ghosh, P., Das, M. T. &Thakur, I. S. 2014. Invitrotoxicityevaluationoforganic extract of landfill soil and its detoxification by indigenous pyrene-degrading Bacillus sp. ISTPY1. *International Biodeterioration & Biodegradation*, 90: 145–151. https: //doi. org/10. 1016/j.

ibiod. 2014. 03. 001.

Sweetman, A. J., Thomas, G. O. & Jones, K. C. 1999. Modelling the fate and behaviour of lipophilic organic contaminants in lactating dairy cows. *Environmental Pollution*, 104(2): 261–270. https: //doi. org/10. 1016/S0269-7491(98)00177-8.

Syers, J. K., Johnson, A. E. &Curtin, D. 2008. *Efficiencyofsoilandfertilizerphosphorus use*. FAO Fertilizer and Plant Nutrition Bulletin No. 18. Rome, Italy, Food and Agriculture Organization of the United Nations. 108 pp. (also available at http: // www. fao. org/docrep/010/a1595e/a1595e00. htm).

Szasz, F. M. 1995. TheImpactofWorldWarIIontheLand: GruinardIsland, Scotland, and Trinity Site, New Mexico as Case Studies. *Environmental History Review*, 19(4): 15–30. https: //doi. org/10. 2307/3984690.

Tang, X., Li, Q., Wu, M., Lin, L. &Scholz, M. 2016. Reviewofremediationpractices regarding cadmium-enriched farmland soil with particular reference to China. *Journal of Environmental Management*, 181: 646–662. https: //doi. org/10. 1016/j. jenvman. 2016. 08. 043.

Tarazona, J. V. 2014. Pollution, Soil. *Encyclopediaof Toxicology*, pp. 1019–1023. Elsevier. (also available athttp: //linkinghub. elsevier. com/retrieve/pii/B9780123864543005315).

Teh, T., Nik Norulaini, N. A. R., Shahadat, M., Wong, Y. &MohdOmar, A. K. 2016. Risk Assessment of Metal Contamination in Soil and Groundwater in Asia: A Review of Recent Trends as well as Existing Environmental Laws and Regulations. *Pedosphere*, 26(4): 431–450. https: //doi. org/10. 1016/S1002-0160(15)60055-8.

Thiele, S. &Brümmer, G. W. 2002. Bioformationofpolycyclicaromatichydrocarbons in soil under oxygen deficient conditions. *Soil Biology* and *Biochemistry*, 34(5): 733– 735. https: //doi. org/10. 1016/S0038-0717(01)00204-8.

Thomas, C. M. &Nielsen, K. M. 2005. MechanismsofandBarriersto, HorizontalGene Transfer between Bacteria. *Nature Reviews Microbiology*, 3(9): 711–721. https: //doi. org/10. 1038/nrmicro1234.

Thompson, R. C. 2004. LostatSea: WhereIsAllthePlastic?*Science*, 304(5672): 838–838. https: //doi. org/10. 1126/science. 1094559.

Tian, W., Wang, L., Li, Y., Zhuang, K., Li, G., Zhang, J., Xiao, X. & xi, Y. 2015. Responsesof microbial activity, abundance, and community in wheat soil after three years of heavy fertilization with manure-based compost and inorganic nitrogen. *Agriculture, Ecosystems&Environment*, 213: 219–227. https: //doi. org/10. 1016/j. agee. 2015. 08. 009.

Tien, Y. -C., Li, B., Zhang, T., Scott, A., Murray, R., Sabourin, L., Marti, R. &Topp, E. 2017. Impact of dairy manure pre-application treatment on manure composition, soil dynamics of antibiotic resistance genes, and abundance of antibiotic-resistance genesonvegetablesatharvest.

Science of The Total Environment, 581–582: 32–39. https: //doi. org/10. 1016/j. scitotenv. 2016. 12. 138.

Tilman, D., Cassman, K. G., Matson, P. A., Naylor, R. & Polasky, S. 2002. Agricultural sustainability and intensive production practices. *Nature*, 418(6898): 671–677. https: // doi. org/10. 1038/nature01014.

Topp, E. 2003. Bacteria in agricultural soils: Diversity, role and future perspectives. *Canadian Journal of Soil Science*, 83(Special Issue): 303–309. https: //doi. org/10. 4141/ S01-065.

Torrent, J., Barberis, E. & Gil-Sotres, F. 2007. Agriculture as a source of phosphorus for eutrophication in southern Europe. *Soil Use* and *Management*, 23(s1): 25–35. https: // doi. org/10. 1111/j. 1475-2743. 2007. 00122. x.

Tőzsér, D., Magura, T. & Simon, E. 2017. Heavy metal uptake by plant parts of willow species: A meta-analysis. *Journal of Hazardous Materials*, 336: 101–109. https: //doi. org/10. 1016/j. jhazmat. 2017. 03. 068.

Tran, B. C., Teil, M. -J., Blanchard, M., Alliot, F. & Chevreuil, M. 2015. Fate of phthalates and BPA in agricultural and non-agricultural soils of the Paris area (France). *Environmental Science and Pollution Research*, 22(14): 11118–11126. https: //doi. org/10. 1007/ s11356-015-4178-3.

Trendel, J. M., Lohmann, F., Kintzinger, J. P., Albrecht, P., Chiarone, A., Riche, C., Cesario, M., Guilhem, J. & Pascard, C. 1989. Identification of des-A-triterpenoid hydrocarbons occurring in surface sediments. *Tetrahedron*, 45(14): 4457–4470. https: // doi. org/10. 1016/S0040-4020(01)89081-5.

Turick, C. E., Knox, A. S. & Kuhne, W. W. 2013. Radioactive elements in soil. Interactions, health risks, remediation, and monitoring. *Soils* and *human health*, pp. 137–154. Boca Raton, Fla, CRC Press.

TwEPA. 2014. Taiwan Toxic Chemical Substance Control Act (TCSCA). [Cited 3 April 2018]. http: // www. chemsafetypro. com/Topics/Taiwan/Taiwan_Toxic_Chemical_Substance_Control_Act_TCSCA. html.

Ulrich, A. E., Schnug, E., Prasser, H. -M. & Frossard, E. 2014. Uranium endowments in phosphate rock. *Science of The Total Environment*, 478: 226–234. https: //doi. org/10. 1016/j. scitotenv. 2014. 01. 069.

UN. 2016. Political declaration of the high-level meeting of the General Assembly on antimicrobial resistance. [Cited 4 April 2018]. http: //www. un. org/en/ga/search/view_doc. asp?symbol=A/RES/71/3.

UNECE. 2011. Globally Harmonized System of classification and labelling of chemicals (GHS). New York and Geneva, United Nations. (also available at https: // www. unece. org/fileadmin/DAM/trans/danger/publi/ghs/ghs_rev04/English/ST- SG-AC10-30-Rev4e. pdf).

UNEP. 1998. RotterdanConventiononthePriorInformedConsentProcedurefor Certain Hazardous chemicals and Pesticides in International Trade. [Cited 4 April 2018]. http: //www. pic. int/TheConvention/Overview/TextoftheConvention/tabid/1048/language/en-US/Default. aspx.

UNEP. 2001. The Stockholm Convention on Persistent Organic Pollutants as amended in 2009. [Cited 4 April 2018]. http: //chm. pops. int/TheConvention/ Overview/TextoftheConvention/tabid/2232/Default. aspx.

UNEP. 2018. Worldcommitstopollution-freeplanetatenvironmentsummit. In: *UN Environment* [online]. [Cited 4 April 2018]. http: //www. unenvironment. org/news-and-stories/press-release/world-commits-pollution-free-planet-environment- summit.

UNFCCC. 2015. TheParisAgreement-mainpage. In: *UnitedNationsFramework Convention on Climate Change* [online]. [Cited 4 April 2018]. http: //unfccc. int/paris_ agreement/items/9485. php.

USAgencyforToxicSubtancesandDiseaseRegistry. 2011. PublicHealthAssessment: U. S. Smelter and Lead Refinery. East Chigaco, Indiana, USS Lead. (also available at https: //www. atsdr. cdc. gov/hac/pha/ussmelterandleadrefinery/ ussleadphablue01272011. pdf).

USEPA. 1984. Method610: PolynuclearAromaticHydrocarbons., p. 25. Cincinati, Environmental Monitoring and SupportLaboratory.

US EPA. 1986. Guidelines for the health risk assessment of chemical mixtures. No. EPA/630/R-98/002. Washington. (also available at https: //www. epa. gov/sites/ production/files/2014-11/documents/chem_mix_1986. pdf).

USEPA. 1998. Method3051A. Microwaveassistedaciddigestionofsediments, sludges, soils, and oils. Washington. (also available at https: //www. epa. gov/sites/production/ files/2015-12/documents/3051a. pdf).

USEPA. 2012. PhthalatesActionPlan., p. 16. (alsoavailableathttps: //www. epa. gov/sites/production/files/2015-09/documents/phthalates_actionplan_revised_2012-03-14. pdf).

US EPA. 2013. Protecting and restoring land: Making a visible difference in communities: OSWER FY13 end of year accomplishments report., p. 47. (also available at https: //www. epa. gov/sites/production/files/2014-03/documents/oswer_ fy13_accomplishment. pdf).

US EPA. 2014a. Sampling and Analysis Plan. Field sampling Plan and Quality Assurance Project Plan. No. R9QA/009. 1. Washington. (also available at https: // www. epa. gov/sites/production/files/2015-06/documents/sap-general. pdf).

USEPA, O. 2014b. PersistentOrganicPollutants: AGlobalIssue, AGlobalResponse. In: *US EPA* [online]. [Cited 4 April 2018]. https: //www. epa. gov/international- cooperation/persistent-organic-pollutants-global-issue-global-response.

USEPA, O. 2014c. ResearchonPer-andPolyfluoroalkylSubstances(PFAS). In: *US EPA*[online].

[Cited 4 April 2018]. https: //www. epa. gov/chemical-research/research- and-polyfluoroalkyl-substances-pfas.

US Federal Register. 1993. 40 CFR Part 503: Standards for the use and disposal of sewage sludge. In: *LII / Legal Information Institute* [online]. [Cited 4 April 2018]. https: //www. law. cornell. edu/cfr/text/40/part-503.

Uzen, N. 2016. Use of wastewater for agricultural irrigation and infectious diseases. Diyarbakir example. *Journal of Environmental Protection and Ecology*, 17(2): 488–497.

Van den Berg, M., Kypke, K., Kotz, A., Tritscher, A., Lee, S. Y., Magulova, K., Fiedler, H. & Malisch, R. 2017. WHO/UNEP global surveys of PCDDs, PCDFs, PCBs and DDTs in human milk and benefit–risk evaluation of breastfeeding. *Archives of Toxicology*, 91(1): 83–96. https: //doi. org/10. 1007/s00204-016-1802-z.

Van der Putten, W. H., Jeffery, S., European Commission, Joint Research Centre & Institute for Environment and Sustainability. 2011. *Soil borne human diseases*. Luxembourg, Publications Office.

Van Kauwenbergh, S. J. 2010. *World Phosphate Rock Reserves* and *Resources*. Alabama, US, International Fertilizer Development Center (IFDC). (also available at https: // pdf. usaid. gov/pdf_docs/Pnadw835. PDF).

Van Maele-Fabry, G., Lantin, A. -C., Hoet, P. & Lison, D. 2010. Childhood leukaemia and parental occupational exposure to pesticides: a systematic review and meta- analysis. *Cancer Causes & Control*, 21(6): 787–809. https: //doi. org/10. 1007/s10552-010- 9516-7.

Vandenhove, H. & Turcanu, C. 2011. Agricultural land management options following large-scale environmental contamination. *Integrated Environmental Assessment* and *Management*, 7(3): 385–387. https: //doi. org/10. 1002/ieam. 234.

Vasseur, P. & Cossu-Leguille, C. 2006. Linking molecular interactions to consequent effects of persistent organic pollutants (POPs) upon populations. *Chemosphere*, 62(7): 1033–1042. https: //doi. org/10. 1016/j. chemosphere. 2005. 05. 043.

Venuti, A., Alfonsi, L. & Cavallo, A. 2016. Anthropogenic pollutants on topsoils along a section of the Salaria state road, central Italy. *Annals of Geophysics*(5). https: //doi. org/10. 4401/ag-7021.

Viret, O., Siegfried, W., Holliger, E. & Raisigl, U. 2003. Comparison of spray deposits and efficacy against powdery mildew of aerial and ground-based spraying equipment in viticulture. *Crop Protection*, 22(8): 1023–1032. https: //doi. org/10. 1016/S0261- 2194(03)00119-4.

Vitousek, P. M., Naylor, R., Crews, T., David, M. B., Drinkwater, L. E., Holland, E., Johnes, P. J., Katzenberger, J., Martinelli, L. A., Matson, P. A., Nziguheba, G., Ojima, D., Palm, C. A., Robertson, G. P., Sanchez, P. A., Townsend, A. R. & Zhang, F. S. 2009. Nutrient Imbalances in Agricultural Development. *Science*, 324(5934): 1519–1520. https: //doi. org/10. 1126/ science. 1170261.

Vollaro, M., Galioto, F. & Viaggi, D. 2017. Thecirculareconomyandagriculture: new opportunities for re-using Phosphorus as fertilizer. Bio-based and Applied Economics.

Volpe, M. G., LaCara, F., Volpe, F., DeMattia, A., Serino, V., Petitto, F., Zavalloni, C., Limone, F., Pellecchia, R., DePrisco, P. P. & DiStasio, M. 2009. Heavymetaluptakein the enological food chain. *Food Chemistry*, 117(3): 553–560. https: //doi. org/10. 1016/j. foodchem. 2009. 04. 033.

Wagner, G. J. 1993. AccumulationofCadmiuminCropPlantsandItsConsequences to Human Health. *Advances in Agronomy*, pp. 173–212. Elsevier. (also available at http: //linkinghub. elsevier. com/retrieve/pii/S0065211308605933).

Wales, A. & Davies, R. 2015. Co-SelectionofResistancetoAntibiotics, Biocidesand Heavy Metals, and Its Relevance to Foodborne Pathogens. *Antibiotics*, 4(4): 567–604. https: //doi. org/10. 3390/antibiotics4040567.

Wallova, G., Kandler, N. & Wallner, G. 2012. Monitoringofradionuclidesinsoilandbone samples from Austria. *Journal of Environmental Radioactivity*, 107: 44–50. https: //doi. org/10. 1016/j. jenvrad. 2011. 12. 007.

Walsh, S.., Maillard, J. -Y., Russell, A.., Catrenich, C.., Charbonneau, D.. & Bartolo, R.. 2003. Development of bacterial resistance to several biocides and effects on antibiotic susceptibility. *JournalofHospitalInfection*, 55(2): 98–107. https: //doi. org/10. 1016/S0195- 6701(03)00240-8.

Walters, E., McClellan, K. & Halden, R. U. 2010. Occurrenceandlossoverthreeyears of 72 pharmaceuticals and personal care products from biosolids–soil mixtures in outdoor mesocosms. *Water Research*, 44(20): 6011–6020. https: //doi. org/10. 1016/j. watres. 2010. 07. 051.

Wan, X., Lei, M. & Chen, T. 2016. Cost–benefit calculation of phytoremediation technology for heavy-metal-contaminated soil. *Science of The Total Environment*, 563–564: 796–802. https: //doi. org/10. 1016/j. scitotenv. 2015. 12. 080.

Wang, F., Wang, Z., Kou, C., Ma, Z. & Zhao, D. 2016. ResponsesofWheatYield, Macro-and Micro-Nutrients, and Heavy Metals in Soil and Wheat following the Application of ManureCompostontheNorthChinaPlain. *PLOSONE*, 11(1): e0146453. https: //doi. org/10. 1371/journal. pone. 0146453.

Wang, S. & He, J. 2013. DechlorinationofCommercialPCBsandOtherMultiple Halogenated Compounds by a Sediment-Free Culture Containing *Dehalococcoides* and *Dehalobacter*. *Environmental Science & Technology*: 130904143020001. https: //doi. org/10. 1021/es4017624.

Wang, T., Wang, Y., Liao, C., Cai, Y. & Jiang, G. 2009. PerspectivesontheInclusionof Perfluorooctane Sulfonate into the Stockholm Convention on Persistent Organic Pollutants[1]. *Environmental Science & Technology*, 43(14): 5171–5175. https: //doi. org/10. 1021/es900464a.

Wang, Z., Li, J., Zhao, J. & Xing, B. 2011. ToxicityandInternalizationofCuONanoparticles to Prokaryotic Alga *Microcystis aeruginosa* as Affected by Dissolved Organic Matter. *Environmental Science*

&*Technology*, 45(14): 6032–6040. https: //doi. org/10. 1021/es2010573.

Wania, F. &MacKay, D. 1996. PeerReviewed: TrackingtheDistributionofPersistent Organic Pollutants. *Environmental Science &Technology*, 30(9): 390A-396A. https: // doi. org/10. 1021/es962399q.

Watkinson, A. J., Murby, E. J., Kolpin, D. W. &Costanzo, S. D. 2009. Theoccurrenceof antibiotics in an urban watershed: From wastewater to drinking water. *Science of The TotalEnvironment*, 407(8): 2711–2723. https: //doi. org/10. 1016/j. scitotenv. 2008. 11. 059.

Watson, A. P. &Griffin, G. D. 1992. Toxicityofvesicantagentsscheduledfordestruction bytheChemicalStockpileDisposalProgram. *EnvironmentalHealthPerspectives*, 98: 259–280.

Wauchope, R. D., Yeh, S., Linders, J. B. H. J., Kloskowski, R., Tanaka, K., Rubin, B., Katayama, A., Kördel, W., Gerstl, Z., Lane, M. &Unsworth, J. B. 2002. Pesticidesoilsorptionparameters: theory, measurement, uses, limitations andreliability. *Pest Management Science*, 58(5): 419–445. https: //doi. org/10. 1002/ps. 489.

Wawer, M., Magiera, T., Ojha, G., Appel, E., Kusza, G., Hu, S. &Basavaiah, N. 2015. Traffic-Related Pollutants in Roadside Soils of Different Countries in Europe and Asia. *Water, Air, & Soil Pollution*, 226(7). https: //doi. org/10. 1007/s11270-015-2483-6.

Welch, R. M., Hart, J. J., Norvell, W. A., Sullivan, L. A. &Kochian, L. V. 1999. Effectsofnutrient solution zinc activity on net uptake, translocation, and root export of cadmium and zinc by separated sections of intact durum wheat (Triticum turgidum L. var durum) seedlingroots. *PlantandSoil*, 208(2): 243–250. https: //doi. org/10. 1023/A: 1004598228978.

WHO. 1993. *TheWHOrecommendedclassificationofpesticidesbyhazardandguidelines toclassifications1992-1993*. Ginebra, WorldHealthOrganization.

WHO. 2001a. IntegratedRiskAssessment. (alsoavailableathttp: //www. who. int/ipcs/publications/new_issues/ira/en/).

WHO. 2001b. Schistosomiasis and soil-transmitted helminth infections. [Cited 4 April 2018]. http: //www. who. int/neglected_diseases/mediacentre/WHA_54. 19_Eng. pdf.

WHO. 2008. Anthrax in humans and animals. Geneva, World Health Organization. (also available at http: //www. who. int/csr/resources/publications/ AnthraxGuidelines2008/en/).

WHO. 2010. PreventingDiseaseThroughHealthyEnvironments. Actionisneeded on chemicals of major public health concern., p. 6. Geneva, World Health Organization.

WHO. 2013. Contaminatedsitesandhealth. Copenhagen, Denmark. (alsoavailable at http: //www. euro. who. int/data/assets/pdf_file/0003/186240/e96843e. pdf).

WHO, ed. 2014. *Antimicrobial resistance: global report on surveillance*. Geneva, Switzerland, World Health Organization. 232 pp.

WHO. 2017a. Soil-transmittedhelminthinfections. In: *WHO*[online]. [Cited4April 2018]. http: //

www.who.int/mediacentre/factsheets/fs366/en/.

WHO. 2017b. Foodsafety. In: *WHO*[online]. [Cited 4 April 2018]. http://www.who.int/mediacentre/factsheets/fs399/en/.

WHO. 2018. Antimicrobialresistance. In: *WHO*[online]. [Cited 4 April 2018]. http://www.who.int/mediacentre/factsheets/fs194/en/.

WHO & FAO. 1995. General Standard for Contaminants and Toxins in Food and Feed. [Cited 4 April 2018]. http://www.fao.org/fileadmin/user_upload/livestockgov/documents/1_CXS_193e.pdf.

Wierzbicka, M., Bemowska-Kałabun, O. & Gworek, B. 2015. Multidimensional evaluation of soil pollution from railway tracks. *Ecotoxicology*, 24(4): 805–822. https://doi.org/10.1007/s10646-015-1426-8.

Wilcke, W. 2007. Global patterns of polycyclic aromatic hydrocarbons (PAHs) in soil. *Geoderma*, 141(3–4): 157–166. https://doi.org/10.1016/j.geoderma.2007.07.007.

Winckler, C. & Grafe, A. 2001. Use of veterinary drugs in intensive animal production: Evidence for persistence of tetracycline in pig slurry. *Journal of Soils* and *Sediments*, 1(2): 66–70. https://doi.org/10.1007/BF02987711.

Wislocka, M., Krawczyk, J., Klink, A. & Morrison, L. 2006. Bioaccumulation of Heavy Metals by Selected Plant Species from Uranium Mining Dumps in the Sudety Mts., Poland. *Polish Journal of Environmental Studies*, 15(5): 811–818.

Withers, P. J. A., Sylvester-Bradley, R., Jones, D. L., Healey, J. R. & Talboys, P. J. 2014. Feed the Crop Not the Soil: Rethinking Phosphorus Management in the Food Chain. *Environmental Science & Technology*, 48(12): 6523–6530. https://doi.org/10.1021/es501670j.

Witte, W. 1998. BIOMEDICINE: Medical Consequences of Antibiotic Use in Agriculture. *Science*, 279(5353): 996–997. https://doi.org/10.1126/science.279.5353.996.

Woywodt, A. & Kiss, A. 2002. Geophagia: The History of Earth-Eating. *Journal of the Royal Society of Medicine*, 95(3): 143–146. https://doi.org/10.1177/014107680209500313.

Wu, C., Spongberg, A. L. & Witter, J. D. 2009. Adsorption and Degradation of Triclosan and Triclocarban in Soils and Biosolids-Amended Soils. *Journal of Agricultural* and *FoodChemistry*, 57(11): 4900–4905. https://doi.org/10.1021/jf900376c.

Wu, C., Spongberg, A. L., Witter, J. D., Fang, M., Ames, A. & Czajkowski, K. P. 2010. Detection of Pharmaceuticals and Personal Care Products in Agricultural Soils Receiving Biosolids Application. *CLEAN-Soil, Air, Water*, 38(3): 230–237. https://doi.org/10.1002/clen.200900263.

Wuana, R. A. & Okieimen, F. E. 2011. Heavy Metals in Contaminated Soils: A Review of Sources, Chemistry, Risks and Best Available Strategies for Remediation. *ISRN Ecology*, 2011: 1–20.

https: //doi. org/10. 5402/2011/402647.

Xia, L., Lam, S. K., Yan, X. &Chen, D. 2017. HowDoesRecyclingofLivestockManure in Agroecosystems Affect Crop Productivity, Reactive Nitrogen Losses, and Soil Carbon Balance? *Environmental Science &Technology*, 51(13): 7450–7457. https: //doi. org/10. 1021/acs. est. 6b06470.

Xia, L., Wang, S. &Yan, X. 2014. Effectsoflong-termstrawincorporationonthenet global warming potential and the net economic benefit in a rice–wheat cropping system in China. *Agriculture, Ecosystems & Environment*, 197: 118–127. https: //doi. org/10. 1016/j. agee. 2014. 08. 001.

Xia, Z., Duan, X., Qiu, W., Liu, D., Wang, B., Tao, S., Jiang, Q., Lu, B., Song, Y. &Hu, X. 2010. Health risk assessment on dietary exposure to polycyclic aromatic hydrocarbons (PAHs) in Taiyuan, China. *Science of The Total Environment*, 408(22): 5331–5337. https: //doi. org/10. 1016/j. scitotenv. 2010. 08. 008.

Xu, M. Y., Wang, P., Sun, Y. -J., Yang, L. &Wu, Y. -J. 2017. Jointtoxicityofchlorpyrifosand cadmium on the oxidative stress and mitochondrial damage in neuronal cells. *Food andChemicalToxicology*, 103: 246–252. https: //doi. org/10. 1016/j. fct. 2017. 03. 013.

Yablokov, A. V., Nesterenko, V. B. &Nesterenko, A. V. 2009. ChapterIII. Consequencesof the Chernobyl Catastrophe for the Environment. *Annals of the New York Academy of Sciences*, 1181(1): 221–286. https: //doi. org/10. 1111/j. 1749-6632. 2009. 04830. x.

Yang, G., Chen, C., Wang, Y., Peng, Q., Zhao, H., Guo, D., Wang, Q. &Qian, Y. 2017a. Mixture toxicity of four commonly used pesticides at different effect levels to the epigeic earthworm, Eisenia fetida. *Ecotoxicology* and *Environmental Safety*, 142: 29–39. https: // doi. org/10. 1016/j. ecoenv. 2017. 03. 037.

Yang, H., Huang, X., Thompson, J. R. &Flower, R. J. 2014. SoilPollution: UrbanBrownfields. *Science*, 344(6185): 691–692. https: //doi. org/10. 1126/science. 344. 6185. 691-b.

Yang, Q., Tian, T., Niu, T. &Wang, P. 2017b. Molecularcharacterizationofantibiotic resistance in cultivable multidrug-resistant bacteria from livestock manure. *EnvironmentalPollution*, 229: 188–198. https: //doi. org/10. 1016/j. envpol. 2017. 05. 073.

Yang, S., Liao, B., Yang, Z., Chai, L. &Li, J. 2016. Revegetationofextremelyacidmine soils based on aided phytostabilization: A case study from southern China. *Science ofTheTotalEnvironment*, 562: 427–434. https: //doi. org/10. 1016/j. scitotenv. 2016. 03. 208.

Yao, H., Xu, J. &Huang, C. 2003. Substrateutilizationpattern, biomassandactivity of microbial communities in a sequence of heavy metal-polluted paddy soils. *Geoderma*, 115(1–2): 139–148. https: //doi. org/10. 1016/S0016-7061(03)00083-1.

Yaron, B., Dror, I. &Berkowitz, B. 2012. *Soil-SubsurfaceChange*. Berlin, Heidelberg, Springer Berlin Heidelberg. (also available at http: //link. springer. com/10. 1007/978- 3-642-24387-5).

Yen, T. -H., Lin-Tan, D. -T. &Lin, J. -L. 2011. Foodsafetyinvolvingingestionoffoodsand beverages prepared with phthalate-plasticizer-containing clouding agents. *Journal of the Formosan Medical Association = Taiwan Yi Zhi*, 110(11): 671–684. https: //doi. org/10. 1016/j. jfma. 2011. 09. 002.

Ying, G. -G., Yu, X. -Y. &Kookana, R. S. 2007. Biologicaldegradationoftriclocarbanand triclosan in a soil under aerobic and anaerobic conditions and comparison with environmental fate modelling. *Environmental Pollution*, 150(3): 300–305. https: //doi. org/10. 1016/j. envpol. 2007. 02. 013.

Yu, Z., Gunn, L., Wall, P. &Fanning, S. 2017. Antimicrobialresistanceanditsassociation with tolerance to heavy metals in agriculture production. *Food Microbiology*, 64: 23– 32. https: //doi. org/10. 1016/j. fm. 2016. 12. 009.

Yuan, Z., Jiang, S., Sheng, H., Liu, X., Hua, H., Liu, X. &Zhang, Y. 2018. HumanPerturbation of the Global Phosphorus Cycle: Changes and Consequences. *Environmental Science &Technology*, 52(5): 2438–2450. https: //doi. org/10. 1021/acs. est. 7b03910.

Zahran, S., Laidlaw, M. A. S., McElmurry, S. P., Filippelli, G. M. &Taylor, M. 2013. LinkingSource and Effect: Resuspended Soil Lead, Air Lead, and Children's Blood Lead Levels in Detroit, Michigan. *Environmental Science &Technology*, 47(6): 2839–2845. https: //doi. org/10. 1021/es303854c.

Zeng, F., Cui, K., Xie, Z., Wu, L., Liu, M., Sun, G., Lin, Y., Luo, D. &Zeng, Z. 2008. Phthalate esters (PAEs): Emerging organic contaminants in agricultural soils in peri-urban areas around Guangzhou, China. *Environmental Pollution*, 156(2): 425–434. https: // doi. org/10. 1016/j. envpol. 2008. 01. 045.

Zeng, X., Li, L. &Mei, X. 2008. HeavyMetalContentinChineseVegetablePlantation LandSoilsa ndRelatedSourceAnalysis. *AgriculturalSciencesinChina*, 7(9): 1115–1126. https: //doi. org/10. 1016/S1671-2927(08)60154-6.

Zhang, H., Luo, Y., Wu, L., Huang, Y. &Christie, P. 2015a. Residuesandpotentialecological risks of veterinary antibiotics in manures and composts associated with protected vegetable farming. *Environmental Science* and *Pollution Research*, 22(8): 5908–5918. https: //doi. org/10. 1007/s11356-014-3731-9.

Zhang, H., Wang, Z., Zhang, Y., Ding, M. &Li, L. 2015b. Identificationoftraffic-related metals and the effects of different environments on their enrichment in roadside soils along the Qinghai–Tibet highway. *Science of The Total Environment*, 521–522: 160–172. https: //doi. org/10. 1016/j. scitotenv. 2015. 03. 054.

Zhanqiang, Q. X. F. 2010. Degradation of Halogenated Organic Compounds by Modified Nano Zero-Valent Iron. *Progress in Chemistry*: Z1.

Zhao, S., Qiu, S., Cao, C., Zheng, C., Zhou, W. &He, P. 2014a. Responsesofsoilproperties, microbial community and crop yields to various rates of nitrogen fertilization in a wheat–maize cropping system in north-central China. *Agriculture, Ecosystems & Environment*, 194: 29–37. https: //doi. org/10. 1016/j. agee. 2014. 05. 006.

Zhao, Y., Yan, Z., Qin, J. &Xiao, Z. 2014b. Effectsoflong-termcattlemanureapplication on soil properties and soil heavy metals in corn seed production inNorthwestChina. *Environmental Science* and *Pollution Research*, 21(12): 7586–7595. https: //doi. org/10. 1007/s11356-014-2671-8.

Zhou, X. &Zhang, Y. 2014. Temporaldynamicsofsoiloxidativeenzymeactivity across a simulated gradient of nitrogen deposition in the Gurbantunggut Desert, Northwestern China. *Geoderma*, 213: 261–267. https: //doi. org/10. 1016/j. geoderma. 2013. 08. 030.

Zhu, J. H., Li, X. L., Christie, P. &Li, J. L. 2005. Environmentalimplicationsoflow nitrogen use efficiency in excessively fertilized hot pepper (Capsicum frutescens L.) cropping systems. *Agriculture, Ecosystems & Environment*, 111(1–4): 70–80. https: // doi. org/10. 1016/j. agee. 2005. 04. 025.

Zhu, Y. G. &Shaw, G. 2000. Soil contamination with radionuclides and potential remediation. *Chemosphere*, 41(1–2): 121–128.

Zouboulis, A. I., Moussas, P. A. &Nriagu, E. -C. J. O. 2011. GroundwaterandSoilPollution: Bioremediation. *Encyclopedia of Environmental Health*, pp. 1037–1044. Elsevier. (also available athttp: //linkinghub. elsevier. com/retrieve/pii/B9780444522726000350).

图书在版编目（CIP）数据

土壤污染：一个隐藏的现实/联合国粮食及农业组织编著；陈保青，刘海涛译．—北京：中国农业出版社，2021.6
（FAO中文出版计划项目丛书）
ISBN 978-7-109-28113-4

Ⅰ.①土⋯ Ⅱ.①联⋯②陈⋯③刘⋯ Ⅲ.①土壤污染－研究 Ⅳ.①X53

中国版本图书馆CIP数据核字（2021）第063865号

著作权合同登记号：图字01-2021-2171号

土壤污染：一个隐藏的现实
TURANG WURAN:YIGE YINCANG DE XIANSHI

中国农业出版社出版
地址：北京市朝阳区麦子店街18号楼
邮编：100125
责任编辑：王秀田　文字编辑：张楚翘
版式设计：王　晨　责任校对：刘丽香
印刷：中农印务有限公司
版次：2021年6月第1版
印次：2021年6月北京第1次印刷
发行：新华书店北京发行所
开本：700mm×1000mm 1/16
印张：8.75
字数：150千字
定价：60.00元

版权所有·侵权必究
凡购买本社图书，如有印装质量问题，我社负责调换。
服务电话：010-59195115　010-59194918